Microorganisms for Sustainability

Volume 38

Series Editor

Naveen Kumar Arora, Environmental Microbiology, School for Environmental Science, Babasaheb Bhimrao Ambedkar University, Lucknow, Uttar Pradesh, India

Microorganisms perform diverse roles on our planet most of which are important to make earth a habitable and sustainable ecosystem. Many properties of microorganisms are being utilized as low input biotechnology to solve various problems related to the environment, food security, nutrition, biodegradation, bioremediation, sustainable agriculture, bioenergy and biofuel, bio-based industries including microbial enzymes/ extremozymes, probiotics etc. The book series covers all the wider aspects and unravels the role of microbes towards achieving a sustainable world. It focuses on various microbial technologies related to sustenance of ecosystems and achieving targets of Sustainable Development Goals. Series brings together content on microbe based technologies for replacing harmful chemicals in agriculture, green alternatives to fossil fuels, use of microorganisms for reclamation of wastelands/ stress affected regions, bioremediation of contaminated habitats, biodegradation purposes. Volumes in the series also focus on the use of microbes for various industrial purposes including enzymes, extremophilic microbes and enzymes, effluent treatment, food products.

The book series is a peer reviewed compendium focused on bringing up contemporary themes related to microbial technology from all parts of the world, at one place for its readers, thereby ascertaining the crucial role of microbes in sustaining the ecosystems.

Ashok Kumar Nadda • Gunjan Goel
Editors

Microbes for Natural Food Additives

Springer

Editors
Ashok Kumar Nadda
Department of Biotechnology & Bioinformatics
Jaypee University of Information Technology
Waknaghat, India

Gunjan Goel
Department of Microbiology
Central University of Haryana
Mahendergarh, Haryana, India

ISSN 2512-1901 ISSN 2512-1898 (electronic)
Microorganisms for Sustainability
ISBN 978-981-19-5713-0 ISBN 978-981-19-5711-6 (eBook)
https://doi.org/10.1007/978-981-19-5711-6

This Springer imprint is published by the registered company Springer Nature Singapore Pte Ltd.
The registered company address is: 152 Beach Road, #21-01/04 Gateway East, Singapore 189721, Singapore

Preface

Taking into consideration the consumer's awareness about health in particular to food additives in foods, the shift towards consumption of natural/microbial source of food additives is the need of the hour. These microbial food additives are gaining interest from consumers and food manufacturing units as well. Although the production of microbial food additives has been shown to have positive health benefits, their application on different food commodities is still limited as there are some drawbacks and limitations with respect to available literature. For the production of microbial food additives, the basic requirements include non-pathogenicity, faster growth kinetics on a cheaper substrate as well as the GRAS status of the microbial source employed. Microbial fermentation changes the complete quality of the raw ingredient through different metabolic pathways. Microbially fermented foods differ from the raw material in terms of its appearance, texture, flavor, and nutritional quality. These natural additives result in technological as well as functional improvement in the product with an overall health benefit. Microbial derived food additives offer a means of producing safe, nutritious foods with desirable organoleptic qualities and extended storage stability. This book will provide the reader the basic concepts of food additives and their different classes. The chapters provide an overview on different categories of food additives such as enzymes, antioxidants, stabilizers, emulsifiers, colorants, and flavoring compounds which are microbial derived products. A focus will also be on the microbes as such added as additive such as probiotics. The book received inputs from renowned collaborators globally to provide unbiased and latest information in the field of food additives. The information is worthy of interest to scientists, researchers, graduate or postgraduate students, and food industries. The reader will get in-depth scientific knowledge on categories of food additives, their production strategies, and health beneficial activities.

The book presents up-to-date information about different categories of microbial derived food additives for their application in the food industry. Alongside the production part, the book also deals with social perspectives and legislative policies regarding the safer use of food additives. The first two chapters introduce the fundamentals of food additives and the role of enzymes in the food industry. Chapters 3 and 4 describe microbial antioxidants and preservatives. Chapters 5 and 6 focus on prebiotic and synbiotic foods and food stabilizers. Various industrial applications of microorganisms in the dairy and food industry are described in Chaps. 7 and 8. The last chapter summarizes the hypersensitivities associated with food additives. We firmly believe that the present book will be beneficial for all the early-stage researchers and industrialists.

Waknaghat, Himachal Pradesh, India — Ashok Kumar Nadda
Mahendergarh, Haryana, India — Gunjan Goel

Contents

Series Editor and Editors

About the Series Editor

Naveen Kumar Arora, Fellow of International Society of Environmental Botanists (FISEB), PhD in Microbiology, is Professor and Head of the Department of Environmental Science at Babasaheb Bhimrao Ambedkar University (A Central University), Lucknow, Uttar Pradesh, India. He is a renowned researcher in the field of environmental microbiology and biotechnology. His specific area of research is plant-microbe interactions, particularly plant growth-promoting rhizobacteria. He has more than 75 research articles published in premium international journals and several articles published in magazines and dailies. He is an editor of 25 books, published by Springer. He is a member of several national and international societies, Secretary General of Society for Environmental Sustainability, serves on the editorial board of four journals, and a reviewer of several international journals. He is also the editor in chief of the journal *Environmental Sustainability* published by Springer Nature. He has delivered lectures at conferences and seminars around the globe. He has a long-standing interest in teaching at the PG level and is involved in taking courses in bacteriology, microbial physiology, environmental microbiology, agriculture microbiology, and industrial microbiology. He has been advisor to 134 postgraduate and 11 doctoral students. He has been awarded for excellence in research by several societies and national and international bodies/organizations. Although an academician and researcher by profession, he has a huge obsession for wildlife and its conservation and has authored a book, *Splendid Wilds*. He is the President of Society for Conservation of Wildlife and has a dedicated website www.naveenarora.co.in for the cause of wildlife and environment conservation.

About the Editors

Ashok Kumar Nadda is working as an Assistant Professor in the Department of Biotechnology and Bioinformatics, Jaypee University of Information Technology, Waknaghat, Solan, Himachal Pradesh, India. He is working in the area of microbial enzyme and biocatalysis for the last 12 years. He has completed his Ph.D. in Biotechnology from the Department of Biotechnology, Himachal Pradesh University, Summer Hill, Shimla, India, in 2013. He worked as a postdoctoral fellow in State Key Laboratory of Agricultural Microbiology, Huazhong Agricultural University, Wuhan, China, for 2 years. He also worked as Brain Pool researcher/Assistant Professor at Konkuk University, Seoul, South Korea. Dr. Ashok has a keen interest in microbial biotechnology, CO_2 conversion, enzyme immobilization, biomass degradation, biofuel synthesis, and biotransformations. Dr. Ashok has published 125 research papers in various internationally reputed journals, 25 books, and 45 book chapters. Dr. Ashok is an editorial member of various journals of international repute.

Gunjan Goel is working as an Professor in the Department of Microbiology, Central University of Haryana, Mahendergarh, India. He has completed his Ph.D. in Microbiology from ICAR-National Dairy Research Institute, Karnal, India. In his doctoral studies, he has focused on analysis of microbial degradation of polyphenols. During his postdoctoral visits to the University of Hohenheim, Germany, and University of Ghent, Belgium, he has worked on the effects of plant secondary metabolites enhancement of functions of rumen microbes as well as metabolic pathways for fatty acids. To his credit, he has published many research papers in journals of repute. Presently, he is working in the area of microbial ecology of gut, bioaccessibility and bioavailability of food ingredients in vitro simulated gut model system, and assessment of probiotics using in vitro and in vivo models. He is a life member of the Association of Microbiologists of India (AMI) and Probiotic Association of India (PAI).

Chapter 1
Microbial Food Additives: Types, Functions, and Challenges

Souparno Paul, Ashok Kumar Nadda, and Gunjan Goel

Abstract The use of food additives enhances organoleptic qualities as well as the appearance and texture of the food. The chemical route of production of these additives poses several health risks and safety concerns; therefore, microbial resources for the production of food additives has attracted the scientific community as well as food industries to counter the demand of consumers towards safer foods. The microbial food additives include flavor compounds, biopreservatives, microbial enzymes, sweeteners, emulsifiers, and others. Various international organizations such as Food and Drug Administration, World Health Organization, and Codex Alimentarius have defined regulations towards the use of food additives, yet there is lack of uniformity in laws regarding their use. Although a lot of advancements have been made in the use of engineered microorganisms for the production of higher levels of food additives during fermentation, however, their stability, toxicity, and microbiological safety need to be assessed during the preservation of foods for a longer period of time.

Keywords Food additives · Regulations · Production of food additives · Risks of food additives

1.1 Introduction

Food additives are part of civilization. For ages, food additives have been added for various reasons and most of the history of how food additives actually came into usage is unknown or not traceable. Spices, which have been used since the early bronze age, are a type of additive and have become part of every cuisine. Salt and

---

S. Paul · G. Goel (✉)
Department of Microbiology, School of Interdisciplinary and Applied Sciences, Central University of Haryana, Mahendergarh, India

A. K. Nadda
Department of Biotechnology and Bioinformatics, Jaypee University of Information Technology, Waknaghat, India

A. K. Nadda, G. Goel (eds.), *Microbes for Natural Food Additives*, Microorganisms for Sustainability 38, https://doi.org/10.1007/978-981-19-5711-6_1

sugar are the most widely used food additives to date. With time the complexity of food processing and cooking and demands for extended shelf life have resulted in the formulation and development of new and at times novel food additives. As food processing changes over geographic and climatic conditions so does the use of food additives. Various diverse kinds of food additives are used with varied functions, and for a long time, it has been a topic of discussion to harmonize its usage (Carocho et al. 2014). As of now, thousands of substances are being used as food additives such as salt, sugars, corn syrups, pigments, and others. As defined by the Food and Drug Administration (FDA), "a food additive is defined as any substance used to provide technical effects in the foods." In recent years, a large number of food industries are using food additives due to increased production of ready-to-eat processed foods. These food additives are used to enhance the flavor, appeal, freshness, and safety of foods. For example, the use of food additives in bread; without any additive it goes stale and hard a day after baking because of the retrogradation of starch due to which it goes back to the crystalline form that it was in the grain before being baked into bread. Also, fungus and molds develop within a very short time. To prevent mold formation, fats are added to give a moist texture to the bread without the addition of water. Another addition is Yeast boost whose sole purpose is to accelerate the dough rising process and increase the production output. Vinegar is an acidulant and works synergistically with a common mold inhibitor, calcium propionate, to inhibit mold growth. Amylase from yeasts acts as a dough conditioner and ascorbic acid to strengthen the dough and to provide elasticity to the gluten. Lecithin in fats acts as an emulsifier that helps water and fats bind together and contributes to the expansion of the dough. Therefore, to make this product with extended shelf life food additives are added. These additives are indispensable today in the food industry where the demand for long shelf-life foods is on the rise.

Food additives are defined as "any substance the intended use of which results or may reasonably be expected to result—directly or indirectly—in its becoming a component or otherwise affecting the characteristics of any food." These additives can be used at different stages of food preparation such as during production, processing, packaging, transportation, and storage of food (Fig. 1.1).

In the context of food additives, a Joint FAO/WHO Expert Committee on Food Additives (JECFA) was constituted in 1956. The objective of this committee was to evaluate the safety regarding the use of food additives. However this committee is also evaluating the safety issues pertaining to presence of other contaminants, naturally occurring toxicants (NOTS), and residues of veterinary drugs in foods. This committee also forms the advisory body to FAO and WHO member governments and to the Codex Alimentarius Commission (CAC). The committee is involved in the development of protocols and principles related to safety assessment of foods. The CAC was formed in 1953 to formulate standards for food safety which are recognized internationally. The aim of this is to protect the health of the consumers and to ensure fair practices in food trade. According to CODEX Alimentarius, "Food additive means any substance, the intentional addition of which to food for a technological purpose, may be reasonably expected to result in it or its by products becoming a component of or otherwise affecting the

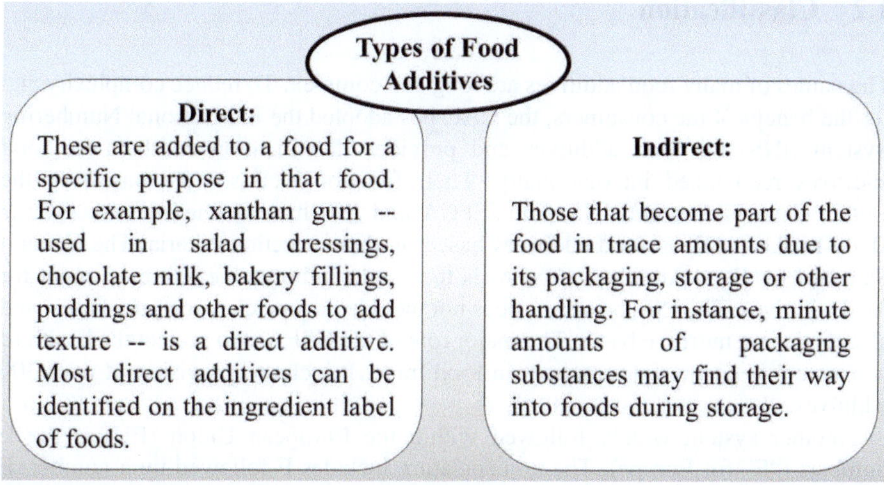

Fig. 1.1 Types of food additives depending on their use

characteristic of such foods." These food additives may or may not have nutritive value, but may add organoleptic qualities in foods without affecting the characteristics of such foods. However, the CAC also explained that the food additive term excludes the presence of other contaminants or substances which are added to food for maintaining or improving nutritional qualities.

Food additives must be used only if required and they do not pose any health hazard to the consumers. Although not restricted to these, food additives are also widely used for the following reasons:

- To extend the shelf life and to enhance or preserve the nutritional quality of the food
- To add specific ingredients or constituents in the foods required for consumers with special dietary needs
- To increase the stability of food or its components or to improve its organoleptic properties

Food ingredients may be "food additives" that are approved by the FDA for specific uses and fall under GRAS (generally recognized as safe) substances. These additives are provided with GRAS status on the basis of recommendations from the experts qualified to evaluate the safety of the substance (FDA 2018). Although there are a plenty of food additives that are permitted to be used in foods under permissible levels, nevertheless, it should always be understood that in any way food additives should not be misused to mask any faulty effects of the raw materials or of undesirable (including unhygienic) practices or techniques or they do not alter the nutritive values or quality of the foods.

1.2 Classification

The names of many food additives are long and complex. To reduce complexity and for the benefit of the consumers, the CAC has adopted the International Numbering System (INS) for food additives and provides identification numbers for food additives recognized internationally. Their functional classes are based on the Codex Class Names and INS. The JEFCA and INS have defined the Acceptable Daily Intake (ADI) of food additives based on their selection criteria. The General Standard for Food Additives (GSFA) is the single authoritative reference point for food additives. The INS, however, does not include flavors, chewing gum bases, and also dietic and nutritive bases. The major role of the GSFA is to protect the health of consumers to ensure fair practices in food trade. It includes provision of over 300 additives.

Another system widely followed within the European Union (EU) is the E numbers ("E" for Europe). The nomenclature includes E followed by a number. E number classifies food additives according to their major application in foodstuff (Kallscheuer 2018).

In the United States, additives are restricted down to a few classes and allow the additives to be mentioned in at least two or more classes. As per recommendations of the FDA, there are more than 3000 food additives distributed into 6 groups which include flavoring agents, preservatives, coloring agents, nutritional additives, texturizing agents, and miscellaneous agents. The FDA has an online database covering "Everything Added to Food in the United States" (EAFUS) which comprises all the compounds added to the food as well as the FDA Redbook which entails all the guidelines and legislation for the food industry (Carocho et al. 2014).

The Food Safety and Standards Authority of India (FSSAI) monitors and sets standards regarding the use of food additives in India. FSSAI provisions are broadly based on JEFCA assessment and guidelines along with their own assessment. Limits are set according to CODEX provisions and also include historically used food additives in the Indian industry. The standards of Identity and Purity are regulated by JEFCA and BIS guidelines (FSSAI 2006).

For application of food additives in foods and to get an overview of how these guidelines are framed, it is important to understand how dosage intake is calculated. For application of an additive for human use, "No Observed Adverse Effect Level (NOAEL) is determined which is the highest dose at which no adverse effects are observed in the most susceptible animal species." The NOAEL value is determined in mg/kg body weight of an individual. Apart from NOAEL, JECFA introduced the term, Acceptable Daily Intake (ADI), which measures the amount of an additive in food that could be ingested orally on a daily basis over a lifetime without an appreciable health risk (Lu 1988; Carocho et al. 2014).

Food additives are broadly grouped into 27 functional classes according to their technological functions. A concise list of them has been compiled below in Table 1.1 (FAO and WHO 2019).

Table 1.1 Functional classes of food additives (FAO and WHO 2019)

Sr. no.	Functional class	Usage	Example	
			INS	
1	Acidity regulator	Controls the acidity or alkalinity of a food	260	• Acetic acid, glacial
			300	• L-Ascorbic acid
			170 (i)	• Calcium carbonate
			501 (i)	• Potassium carbonate
			262 (i)	• Sodium acetate
2	Anti-caking agent	Reduces the aggregation of particles of food	170 (i)	• Calcium carbonate
			504 (i)	• Magnesium carbonate
			421	• Mannitol
3	Antifoaming agent	Prevents or reduces foaming	404	• Calcium alginate
			471	• Mono- and di-glycerides of fatty acids
4	Antioxidants	Extend shelf life of the food by protecting oxidation of food components	300	• L-Ascorbic acid
			320	• Butylated hydroxyanisole (BHA)
			330	• Citric acid
5	Bleaching agent	For decolorization of foods. Bleaching agents do not include pigments	928	• Benzoyl peroxide
			221	• Sodium sulfite
			220	• Sulfur dioxide
6	Bulking agent	Contributes to the bulk of a food	420 (i)	• Sorbitol
			421	• Mannitol
			414	• Gum arabic (Acacia gum)
7	Carbonating agent	Provides carbonation in a food	290	• Carbon dioxide
8	Carrier	Used to deliver food additive via dissolution, dilution, dispersion of other physical methods without any alteration of function of food additive	901	• Beeswax
			462	• Ethyl cellulose
9	Color	Adds or restores color in a food	102	• Tartrazine
			101 (iii)	• Riboflavin from *Bacillus subtilis*
			150a	• Caramel I
10	Color retention agent	Stabilizes, retains, or intensifies the color of a food	330	• Citric acid
			504 (i)	• Magnesium carbonate
			511	• Magnesium chloride

(continued)

Table 1.1 (continued)

Sr. no.	Functional class	Usage	Example	
			INS	
11	Emulsifier	For uniform emulsion of two or more phases in a food	1400	• Dextrins, roasted starch
			440	• Pectins
			415	• Xanthan gum
12	Emulsifying salt	For rearrangement of proteins in order to prevent fat separation	327	• Calcium lactate
			500 (i)	• Sodium carbonate
			437	• Tamarind seed polysaccharide
13	Firming agent	To maintain firmness and crispiness in tissues of fruit or vegetables, to strengthen the gel	520	• Aluminum sulfate
			509	• Calcium chloride
			466	• Sodium carboxymethyl cellulose (cellulose gum)
14	Flavor enhancer	Enhances the organoleptic qualities of foods	1104	• Lipases
			1101 (ii)	• Papain
			508	• Potassium chloride
15	Flour treatment agent	For improvement of baking quality or color of dough	1100 (ii)	• alpha-Amylase from *Bacillus stearothermophilus*
			1100 (vi)	• Carbohydrase from *Bacillus licheniformis*
			1101 (i)	• Protease from *Aspergillus oryzae* var.
16	Foaming agent	To maintain a uniform dispersion of a gaseous phase in a liquid or solid food	400	• Alginic acid
			415	• Xanthan gum
			941	• Nitrogen
17	Gelling agent	Provides texture through formation of a gel	418	• Gellan gum
			440	• Pectins
			401	• Sodium alginate
18	Glazing agent	Provides a shiny appearance or provides a protective coating	1503	• Castor oil
			461	• Methylcelluloses
			1200	• Polydextrose
19	Humectant	Prevents food from drying out	422	• Glycerol
			421	• Mannitol
			1200	• Polydextrose
20	Packaging gas	For preservation of foods from oxidation or spoilage	290	• Carbon dioxide
			941	• Nitrogen
			942	• Nitrous oxide
21	Preservative	Extension of shelf life of foods	210	• Benzoic acid
			1105	• Lysozyme
			290	• Carbon dioxide

(continued)

Table 1.1 (continued)

Sr. no.	Functional class	Usage	Example INS	
22	Propellant	Expels a food from a container	290	• Carbon dioxide
			941	• Nitrogen
			942	• Nitrous oxide
23	Raising agent	For increasing volume of dough or batter	342 (i)	• Ammonium dihydrogen phosphate
			341 (ii)	• Calcium hydrogen phosphate
			500 (i)	• Sodium carbonate
24	Sequestrant	Controls the availability of a cation	330	• Citric acid
			401	• Sodium alginate
			420 (i)	• Sorbitol
25	Stabilizer	For maintenance of uniform dispersion of two or more component	404	• Calcium alginate
			1103	• Invertases
			1200	• Polydextrose
26	Sweetener	Imparts a sweet taste to food	421	• Mannitol
			420 (i)	• Sorbitol
			953	• Isomalt
27	Thickener	Increases the viscosity of food	462	• Ethyl cellulose
			422	• Glycerol
			419	• Gum ghatti

1.3 Microbial Production of Food Additives

Food additives can be produced via synthetic or biological ways. The synthetic or the chemical route of production of additives offers few advantages such as low cost and well-established scientific methods of production; however, all these methods have severe disadvantages as the production process might include use of toxic or hazardous chemical compounds, use of extreme physical conditions for production such as heat, pressure, and others, and production of stereospecific chemicals. All these synthetic compounds may offer toxicity after ingestion or during digestion and fermentation in the human gut. To counter these drawbacks, the microbial production of food additives is competitively advantageous of being a viable economical alternative for the production of numerous value-added food ingredients. The microorganisms or their metabolites offer several advantages such as being natural food additives, fast growth kinetics of microorganisms, no use of toxic compounds for extraction of additives, and complete understanding of metabolic pathways of an organism which can be used to alter the titer levels of the metabolite/additive. In comparison to chemical synthesis and production of food additives, microbial production of enzymes and food additives is clean, eco-friendly, and stable in extreme conditions of organic solvents, pH, and temperature. Genetic modification and manipulation of microbial strains have gained prominence as these genetic engineering processes provide exemplary and desired yields yet they can be grouped under the "natural" category. For industrial production of microbial food additives, engineered *Escherichia coli* strains, *Saccharomyces cerevisiae, and Lactobacillus* spp. are commonly used (Sun et al. 2021).

In principle, there are three sources of the compounds which can be used as food additives:

1. The active compounds can be directly extracted from the natural producer (e.g., plant material).
2. The compounds may be synthesized using chemical reactions.
3. The genes responsible for active ingredient can be cloned in a suitable host and the additive can be produced from this engineered microorganism.

Renin (Chymosin) from the calf stomach was extracted by Christian Hansen to manufacture cheese. This marked the beginning of the industrial production of enzymes in food processing.

Apart from the list provided in Table 1.2, there are many other food additives such as xanthan which is an extracellular polysaccharide (EPS) material secreted by microorganisms. Chemically, xanthan is a hetero-polysaccharide with linked glucuronic and mannose units. For industrial applications, xanthan is produced by aerobic fermentation using *Xanthomonas campestris* with an immobilized fermentation process (Jindal and Singh Khattar 2018). Another food additive in this category is Erythritol, used as food sweetener, and is produced by various fungal and mold species (Sun and Alper 2020). L-Glutamate is widely used for its umami taste which is produced from *Corynebacterium glutamicum* (Sano 2009). Riboflavin (vitamin B2), an approved yellow food colorant and dietary supplement, is produced from *Ashbya gossypii* or genetically engineered *Bacillus subtilis* strains (Liu et al. 2020).

Antimicrobial oligopeptide like Nisin is the only bacteriocin approved for commercial use as food preservative. It is produced from *Lactococcus lactis* in whey followed by direct recovery from fermentation broth with foaming fractionation (Özel et al. 2018). There are many other important food additives produced from microorganisms; however it is difficult to add all the details here. The list of food additives is nonexhaustive; a major category of food additive includes sweeteners, flavor compounds, pigments, emulsifiers, and enzymes which have been discussed in other chapters of the book.

Table 1.2 A compiled list of food additives derived from microorganisms (FDA 2018). The list does not contain all the microbial food additives

Sr. no.	Food additive	Microorganism species used	Function
1	Natamycin	*Streptomyces natalensis* and *Streptomyces chattanoogensis*	For extended shelf life of foods
2	Bakers yeast protein	*Saccharomyces cerevisiae*	For baking purposes
3	Gibberellic acid	*Fusarium moniliforme*	Malting of barley
4	Dried yeasts	*Saccharomyces cerevisiae, Saccharomyces fragilis*	For baking purpose
5	Dried torula yeast	*Candida utilis*	For baking purpose
6	Bakers yeast glycan	*Saccharomyces cerevisiae*	For baking purpose
7	Amyloglucosidase	*Rhizopus niveus*	Depolymerization of starch
8	Carbohydrase and cellulase	*Aspergillus niger*	Processing of shrimps
9	Carbohydrase	*Rhizopus oryzae*	Conversion of starch to dextrose
10	Catalase	*Micrococcus lysodeikticus*	Cheese manufacture
11	Esterase-lipase	*Mucor miehei* var. Cooney et Emerson	Used as flavor enhancer in cheeses, fats, and oils, and milk products
12	Alpha-galactosidase	*Mortierella vinaceae var. raffinoseutilize*	Conversion of beet sugar into sucrose
13	Milk-clotting enzymes	*Endothia parasitica, Bacillus cereus, Mucor pusillus Lindt* and *Mucor miehei* and *Aspergillus oryzae* modified to contain the gene for aspartic proteinase *from Rhizomucor miehei* var. Cooney et Emerson	In cheese production
14	Production of citric acid	*Candida guilliermondii*	Flavoring agent and preservative
		Candida lipolytica	

1.4 Production Strategies for Food Additives

In order to obtain efficient and large-scale biosynthesis of food additives and colorants (especially those that are traditionally extracted from plants), there are several intermediate steps. A concise explanation is provided in Fig. 1.2. With the technological advancements in the area of molecular biology and genetic engineering, new ways have been opened to use genetically engineered microorganisms to produce food additives at higher scale. This technology has the potential to get access to an even broader range of accessible products. The first step is the high-throughput screening of the genes responsible for the biosynthesis of an additive. Screening using Omic technologies provides a platform to screen multiple compounds in any plant source. Followed by screening, next step is cloning the gene of

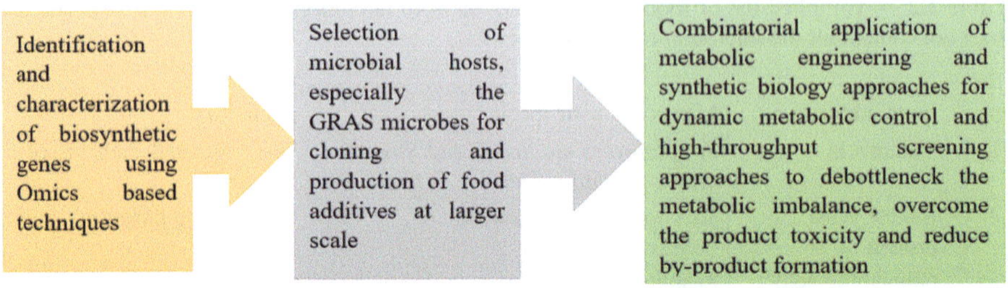

Fig. 1.2 Production strategy of a food additive of plant origin in microbial strains

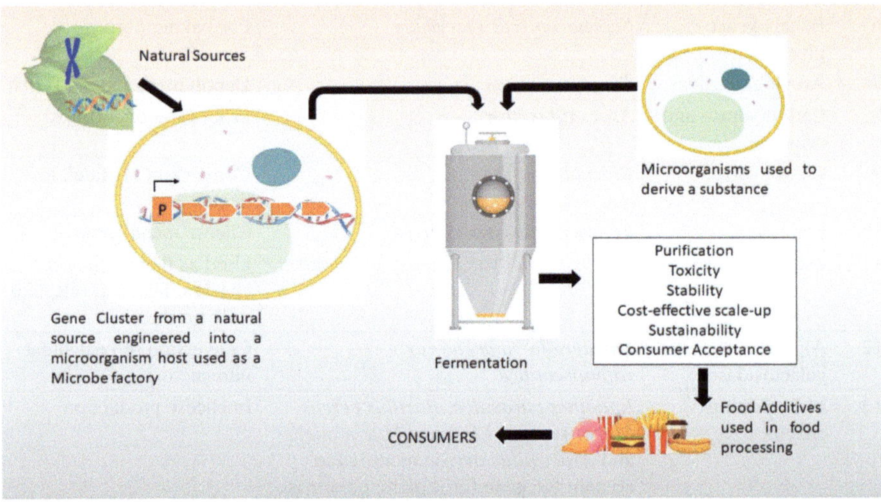

Fig. 1.3 A general flowchart of production methods and application of microbial food additives

interest in a suitable prokaryotic or eucaryotic host to express the food additive in the microbial cell. However, the host strain should be nonpathogenic, should have faster growth rate, could be propagated on cheaper media, and should produce higher biomass. This step involves the optimization of fermentation conditions for the maximum productivity of the compound of interest. Followed by the secretion in the spent medium, further combinatorial approaches are used to limit the toxicity of the compounds and reduction in production of by-products during the fermentation. This step involves the application of metabolic engineering coupled with synthetic biology approaches to deduce the linkages between the metabolic pathways with the production levels.

The production of food additive from a natural plant origin or microbial origin is detailed in Fig. 1.3. In both the approaches, as stated above after the expression of cloned gene is suitable host such as *E. coli, S. cerevisiae, Pichia pastoris, Y. lipolytica* and others. The additives produced from native microflora can be used as such. In both cases, the native and engineered microorganisms are subjected to optimization of bioreactor conditions for maximum productivity of the food additives. The key factors to be taken into consideration include:

(a) Downstream processing: The food additive from the spent broth should be purified using standard protocols and characterized for its purity.
(b) Toxicity: The additive should be checked for its toxicity using in vitro or in vivo models.
(c) Stability: The stability of additive should be checked in different food matrices and different food processing conditions, so that it should not get degraded or transformed in other compounds during food processing conditions.
(d) Consumer acceptance: As consumers are now much aware of health concerns regarding addition of any additives in foods, therefore the food additive in concern should be very well accepted by the consumers.

1.5 Conclusions and Future Perspectives

The production and use of microbial food additives is becoming an attractive alternate to the chemical methods of production; however it is still filled with various challenges. The primary challenge is the cost of raw materials used for production, which makes the process very expensive. For this alternate natural waste materials need to be looked upon to reduce the cost of production of food additives. Another challenge in use of food additives is regarding the set regulations on permissible levels. Even though regulations are set which allow only the use of permitted compounds/colors, often processing units resort to using prohibited compounds/colors to save the cost. Although researches claim on the stability of microbial food additives however, the improper food storage infrastructure and cold storage often impact the stability of food additives.

With globalization and the easy availability of information, there is a growing awareness among consumers who are opting for "additive-free" and "preservative-free" food options. Brands have embraced "Preservative Free" options and are marketing them aggressively. As it was explained in the beginning of the chapter on how additives are used in each step of bread processing, it should be understood that most of the processed foods have additives added at most of the production steps to ease and economize processing, so in reality majority of the foods have chemical additives in some form or other.

The additives that impart a marketable quality, such as color and flavor enhancers, may cause allergic reactions, hypersensitivity, digestive disorders, and nervous disorders; therefore, the food processing industry should take a step in limiting the use of chemical food additives and to look for microbial food additives under strict restriction, and verification in accordance with national food safety guidelines. Although there are a plethora of microbial food additives still there is a great scope to explore unidentified additives from natural resources. Newer technologies in gene manipulation and synthetic biology approaches will pave the way to produce natural food additives economically. However, before their application in foods, careful studies and validation should be carried out to decide the fate of additives at the levels of legislators, researchers, food industries, and final stakeholders, the consumer.

References

Carocho M, Barreiro MF, Morales P, Ferreira ICFR (2014) Adding molecules to food, pros and cons: a review on synthetic and natural food additives: adding molecules to food, pros and cons.... Compr Rev Food Sci Food Saf 13(4):377–399. https://doi.org/10.1111/1541-4337.12065

FAO and WHO (2019) Codex General Standard for Food Additives (GSFA) online database. Updated up to the 42nd session of the Codex Alimentarius Commission (2019). https://www.fao.org/gsfaonline/reference/techfuncs.html. Accessed 18 Apr 2022

FDA (2018) Microorganisms & microbial-derived ingredients used in food (partial list). https://www.fda.gov/food/food-ingredients-packaging/overview-food-ingredients-additives-colors. Accessed 23 Apr 2022

FSSAI (2006). https://fssai.gov.in/cms/food-safety-and-standards-act-2006.php. Accessed 2 May 2022

Jindal N, Singh Khattar J (2018) Microbial polysaccharides in food industry. In: Biopolymers for food design. Academic, New York, pp 95–123. https://doi.org/10.1016/b978-0-12-811449-0.00004-9

Kallscheuer N (2018) Engineered microorganisms for the production of food additives approved by the European union—a systematic analysis. Front Microbiol 9:1746. https://doi.org/10.3389/fmicb.2018.01746

Liu S, Hu W, Wang Z, Chen T (2020) Production of riboflavin and related cofactors by biotechnological processes. Microb Cell Factories 19(1):31. https://doi.org/10.1186/s12934-020-01302-7

Lu FC (1988) Acceptable daily intake: inception, evolution, and application. Regul Toxicol Pharmacol 8:45–60

Özel B, Şimşek Ö, Akçelik M, Saris PEJ (2018) Innovative approaches to nisin production. Appl Microbiol Biotechnol 102(15):6299–6307. https://doi.org/10.1007/s00253-018-9098-y

Sano C (2009) History of glutamate production. Am J Clin Nutr 90(3):728S–732S. https://doi.org/10.3945/ajcn.2009.27462f

Sun L, Alper HS (2020) Non-conventional hosts for the production of fuels and chemicals. Curr Opin Chem Biol 59:15–22. https://doi.org/10.1016/j.cbpa.2020.03.004

Sun L, Xin F, Alper HS (2021) Bio-synthesis of food additives and colorants—a growing trend in future food. Biotechnol Adv 47:107694. https://doi.org/10.1016/j.biotechadv.2020.107

Chapter 2
Applications of Enzymes in Food Industries as Additives

Uday S. Annapure, Srutee Rout, and Prem Prakash Srivastav

Abstract In the food industry food additives, flavors, and colorants are immensely utilized in order to improve the quality and safety of food throughout storage, processing, and packaging. Additives can also enhance the nutritional content of some products while also improving their texture, taste, color, and consistency. As a result, food additives are now required in all the food products we are consuming as well as in the food industry too. These compounds are generally obtained through one of the three methods: bio-production, extraction using natural sources, or chemically synthesized, with the last two being the most common. Though chemical synthesis is less expensive, the health hazards linked with it are still a worry. Chemical reactions for the metabolism in the living species are usually driven by enzymes. Food additives and raw materials have been processed using biocatalysts since the dawn of time. Although various enzymes have been found to be helpful in the biosynthesis of food additives and flavors, the production of these compounds in industry using enzymes still remain unnoticed. As a result, we have focused on enzymatic production technology being the most important and commercially viable option for the production of a wide range of natural flavors, food additives, and antioxidants in this chapter.

Keywords Food additives · Flavors · Colorants · Antioxidants · Enzymes

U. S. Annapure (✉)
Department of Food Engineering and Technology, Institute of Chemical Technology, Mumbai, India

Institute of Chemical Technology, Jalna, India
e-mail: us.annapure@ictmumbai.edu.in

S. Rout · P. P. Srivastav
Agricultural and Food Engineering Department, Indian Institute of Technology Kharagpur, Kharagpur, India

A. K. Nadda, G. Goel (eds.), *Microbes for Natural Food Additives*, Microorganisms for Sustainability 38, https://doi.org/10.1007/978-981-19-5711-6_2

2.1 Introduction

Chemicals are added to food to keep them fresh and to improve their color, flavor, or texture. Antioxidants, stabilizers, thickeners, and emulsifiers, as well as sweeteners, flavorings, nutritional additives, and colorants are all added to food ingredients or food for retaining the flavor, changing its appearance, or for enhancing taste (Martins et al. 2018). The United States Food and Drug Administration (FDA) defines colorants or color additives as "any dye, pigment, or chemical that, when added to food, drug, or cosmetic or to the human body, is capable of imparting color" (FDA 2020). When these chemicals and colorants are combined, they improve the quality, appearance, and safety of food throughout its whole life cycle, from processing to consumption. Though FAO/WHO Expert Committee on Food Additives (JECFA) evaluates all of these compounds there are some specific rules and regulations that differ from country to country. FDA's EAFUS list (Everything Added to Food in the United States) contains over 3000 food additives and colorants. Around 402 additives are designated under E-number (European Union inventory number) whereas 2000 substances are assigned under the CNS (Chinese Number System) and the NFSSUFA (National Food Safety Standard for Uses of Food Additives).

The food industry is always investigating new technologies to fulfill consumer demand, and enzymes are one of the most important tools for converting raw materials into finished goods. The biocatalytic transformation of purified enzymes is one promising option for the synthesis of food additives. Catalytic reactions are carried out under moderate operating conditions with excellent stereoselectivity (Lin et al. 2009; Akanbi et al. 2012). Traditionally, enzymes were utilized for preparing food for improving the texture and flavor, thus scaling-up production in industry (Agyei et al. 2019). Enzymes are mostly utilized to make wine, cheese, and a variety of fermented foods and pastry products. Novel enzymes with specific characteristics are finding applications in various food products as technology is improving with time. Enzymes are small proteins that help in catalyzing biological reactions but are not consumed during the process. As a result, natural food additives formed are pure and are made without the need of complex separation and purifying methods. Commercial enzymes have been steadily increasing in the food industry over the last few years as protein technology has advanced (Okolie et al. 2019). They are now a major component in the production of fruits, vegetables, meat, milk products as well as alcoholic and nonalcoholic beverages. Extraction of enzymes is mainly carried out from animals, plants, and microorganisms such as bacteria, yeast, actinomycetes, and molds (Raveendran et al. 2018; Akanbi et al. 2020; Li et al. 2018). Many enzymatic processes are free of solvents and eco-friendly (Huang et al. 2020; Bhavsar and Yadav 2019). Researchers favor microbial enzymes over other enzymes because of their stability and hence can be optimized and produced in a cost-effective manner (Gurung et al. 2013). Many novel enzymes are presently utilized to generate a variety of foods in which potentially carcinogenic or hazardous chemicals are replaced by biocatalysts, due to diversification in many physical and chemical properties of protein. Technical enzyme preparations are more economically viable since production costs are reduced and they may be made more environmentally friendly by using renewable resources, which often improves product quality. Furthermore, preservation has a substantial impact on food and beverage quality. Modern methods, for example, convert juices into concentrates that may be stored for a long time without losing quality but losing its aroma. Enzymes help food products maintain their flavor and color throughout storage. Finally, as biotechnology advances, numerous approaches have emerged, offering unpredictable solutions to existing problems as well as opening up new opportunities. Enzymes are suggested as remarkable "green" technology agents since they can be utilized to either treat or prevent the creation of biological wastes. As a result, many refined enzymes are used not just in food processing but also as food additives.

2.2 Enzymes as Food Additives

Flavors and food additives are critical elements of foods because they enhance flavor, color, and texture while also extending shelf life. In the food industry, the enzymes required for manufacturing are modulated as food additives (Sun et al. 2021). These substances are proteinous nontoxic substances found naturally in foods. They may be produced by microorganisms for catalyzing various reactions. Their application in the food industry is based on three main principles: quality control of specific foods, alteration of the attributes of certain food additives, and modification of the foods themselves. Many enzymes are being exploited as food additives in a variety of food applications. By 2023, the global market for food enzymes is estimated to reach $3.23 billion. Food industry is highly dependent on the enzymes for the processing of different food products. Of the industrial enzyme market (25%), food enzymes are considered to be the second largest segment as it includes the enzymes used for baking (modification of dough rheology, increasing of crumb softness, and gas retention), brewing (clarification of wine, juice, and other beverages), and cheese making (curdling of milk) (Agyei et al. 2019).

Amylases are the most important enzymes especially in the baking industry as they help in accelerating yeast fermentation for dough expansion, thus intensifying flavors and crust color due to browning and caramelization reactions. Due to their high thermostability, chemical starch hydrolysis has been fully replaced in the starch processing industry (Pandey et al. 2000). Amylase enzymes are also essential for the manufacturing of corn syrup, along with acids. In the alcohol industry, alpha amylases and glucoamylases are preferred; thus malt has been replaced completely. In the dairy industry, chymosin is used for coagulation of milk proteins during cheese production. Lactase and beta-galactosidase help in the breakdown of milk lactose into glucose and galactose, allowing lactose-intolerant people to absorb the final product. In the meat industry, papain is utilized to tenderize muscle tissues. Pectinase, xylanase, and cellulase are some of the hydrolytic enzymes which increase the expulsion of juice from the pulp by breaking down cell walls for producing fruit juices. The enzymes pectinases and amylases are also administered to clarify juice. Microorganisms used for the manufacturing of enzymes must not leave any residues that are hazardous to human health. Indigenous or mutant strains produced from native strains using methods such as recombinant DNA technology, serial culture and selection or mutagenesis and selection may be used in the manufacturing of enzyme preparations. Several fungal species are known to have strains that are able to produce low quantities of mycotoxins under fermentation conditions, despite knowing that nonpathogenic and nontoxigenic microorganisms are generally used for enzyme manufacturing in the food industry. Enzyme preparations made from these fungal species should not include toxicologically relevant quantities of mycotoxins, which these species may produce. Because no precise knowledge of substrate or enzyme activity was available, the first generations of enzyme preparations were unspecific. Manufacturers, on the other hand, have been consistently exploring the technical enzymes specificity. Extracellular enzymes are preferable over intracellular enzymes because the recovery and purifying processes are significantly simpler. Another crucial factor is that the organism must be capable of producing large quantities of the necessary enzyme in a suitable time frame. Only a few microbes such as *Trichoderma, Aspergillus, Streptomyces*, and *Bacillus* produce the majority of industrial enzymes. Submerged fermentation (SmF) or liquid surface fermentation (SLF) processes are mostly used by enzyme manufacturers for producing enzymes. As cost of product recovery is inversely proportional to concentration in the fermentation broth, such levels are required if certain chemicals are to be regarded as commodities. According to research and literature survey, there is a significant interest in using solid-state fermentation (SSF) techniques to produce a wide variety of enzymes because it is economic and cost-effective too (Viniegra-González et al. 2003).

Another milestone for scientists in the food industry was the discovery of digestive enzymes. In the 1930s, pectinases were used for improving the pressing quality and clarification of grape juices. Hydrolysis of galactose residues to form galactooligosaccharides (GOS) occurs by glycosidic hydrolase enzyme which is known as galactosidase. This enzyme helps in the elimination of raffinose oligosaccharides in soybean and beetroot processing and is also used as a food additive. Galactosidase degrades lactose during cheese making and also produces prebiotic GOS (Katrolia et al. 2014).

Enzymes intended for use in food must meet the following criteria: (1) should be cost-effective and easily available, (2) should be approved by regulatory bodies, (3) should be efficient at maximum substrate concentrations, (4) optimum temperature and pH must be clearly indicated (Fig. 2.1).

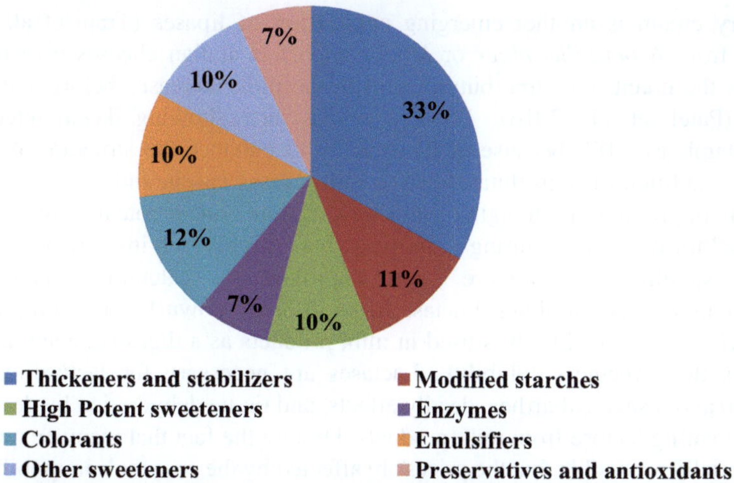

Fig. 2.1 World consumption of food additives. (Source: IHS Markit 2020)

2.3 Application of Enzymes in Various Food Industries

The application of enzymes in different industries is an age-old process. Plants, animals, and microbes are the main sources of enzymes. In the food sector, enzymes are required for the manufacturing of dairy, meat, cereal, and confectionery goods as well as in the beverage and baking industries. Enzymes have been applied in a wide variety of food products to improve quality while reducing overall processing time and cost.

2.3.1 Enzymes Used in Dairy Industry

In the dairy industry, catalase, aminopeptidase, proteases, lipases, lactoperoxidase, and transglutaminase are the principal enzymes used that are obtained from microbes. These enzymes add flavor and improve the quality of a variety of items including curd, cheese, and bread (Abada 2019). Rennet is a well-known exogenous enzyme that has been utilized in dairy processing since 6000 BCE. Rennin causes milk to coagulate by both enzymatic and non-enzymatic activity. Proteases help in significantly improving the flavor of fermented milk products. Proteinases are the first enzymes in the proteolytic enzyme system that breaks down milk protein into peptides. Peptides are broken down into amino acids and tiny peptides using the enzyme peptidases (Abada 2019). When proteinases were combined with peptidases or/and lipases, cheeses formed had exceptional flavor and aroma, thus inhibiting its bitterness levels. Short chain fatty acids responsible for pungent and spicy tastes are released in large portion by animal lipases. Enhancing the flavor and texture of butter and dairy cream is another emerging application of lipases (Trani et al. 2017). Lipases from *Aspergillus niger* or *Mucor miehei* in Italian cheeses give a richer flavor as the quantity of free butyric acid in the milk increases before adding the rennet (Patel et al. 2016). Food scientists are showing keen interest in transglutaminase (TG) because of its capacity to enhance the structure of protein gels. The addition of transglutaminase to milk causes casein and whey proteins to crosslink, improving the strength of milk gels. Casein concentrate and milk proteins were used for the manufacturing of nutrition bars (highly rich in protein) where the storage, stability, and texture were improved by calcium reduction and transglutaminase crosslinking. Lactase helps to break down lactose into galactose and glucose more quickly. It is used in milk products as a digestive agent and also increases the sweetness solubility. Lactases are necessary for lactose-intolerant people to avoid severe diarrhea, deadly effects, and tissue dehydration by decreasing and eliminating lactose from milk products. Despite the fact that many enzymes are commercially accessible, but the prices are affected by the enzyme's immobilization, immobilization method, and carrier type (Abada 2019). Scientists are now making efforts using recombinant engineering techniques for the discovery of new enzymes (Spohner et al. 2015).

2.3.2 *Enzymes Used in Beverage Industry*

Tannases, cellulases, and pectinases are the most important enzymes that are being continuously used in the fruit processing industry. Fruit juice that has been treated with enzymes has numerous advantages over regular processing. These enzymes increase the yield, total soluble acids, and clarification of fruit juice. Apart from this, enzymes lower turbidity and viscosity by increasing the pulp liquefaction (Ramadan 2019). Pectin, proteins, cellulose, tannins, starch, and lignin all contribute to the cloudiness of fruit juice. Macerating enzymes are enzymes that are utilized in the extraction and clarification of fruit juice. Glycoside hydrolases break down glycosidic bonds and are commonly utilized in the manufacturing of syrup, beverages, and bread. Amylases are used in a variety of industries, including brewing and food (Zhang et al. 2018). Before yeast can generate alcohol from barley, starch must be broken down into fermentable sugars. Enzymes such as amylases and proteases are formed during germination whereas some enzymes such as amylases are already present in the barley. The finished malt contains all the enzymes required for the grains to convert into fermentable liquids. The malt enzymes on the other hand have some limitations. However, at a particular pH and temperature, the exogenous enzymes which are tailor-made and available commercially can be used to improve their enzymatic strength. Barley malt is the most common enzyme source to prepare beer from cereals (Meshram et al. 2019). Citrus juices contain bitter chemicals such as naringin, limonin, and neohesperidin. Naringin is the most bitter component having a threshold value for taste of about 20 ppm in water. Grapefruit juice's bitterness has been a key barrier for commercial acceptability for a long time. Some chemical techniques like absorptive bittering, cyclodextrin treatment, and the use of DVB (divinyl benzene styrene) resins all are used to lower naringin levels. Acid hydrolysis for example creates naringenin, a bitter aglycon in addition to rhamnose and glucose from naringin. The use of L-Rhamnosidases allowed for juice debittering (Spohner et al. 2015). Pectinases have sparked widespread attention as a biological catalyst in a variety of industrial hubs. Pectinase catalyzes the depolymerization and de-esterification processes, thus degrading the pectic compounds (hydrolases and lyases). Scientists have also shown keen interest on cellulases too because of their potential to digest cellulosic biomass and convert it to usable products. Cellulases (exoglucanases, endoglucanases, and glucosidases) must work together to depolymerize cellulose so that it can be transformed into valuable compounds by microbes. Tannases (tannin acylhydrolases) are a class of enzymes used in a variety of industrial processes including the production of tea and coffee which are considered the most popular beverages around the world (Ramadan 2019). Cellulases, hemicellulases, galactomannase, and pectinases from *Saccharomyces marscianus, Leuconostoc mesenteroides*, *Flavobacterium* spp., and *Fusarium* spp. are all required for tea and coffee processing.

2.3.3 Enzymes Used in Other Food Industry

Food (baking, dairy products, and starch conversion), textiles, biosensors, health care and nutrition, detergents, pharmaceuticals, and more recently biofuels such as bio-ethanol and bio-diesel are all examples of industrial enzyme applications (Homaei 2016). Softness of meat is influenced by a variety of factors; thus by using enzymes in the meat industry, it will improve the manufacturing process as well as enhance the quality of meat. Researchers have discovered that meat is a rich source of many bioactive peptides having a variety of health-promoting applications. They are made by strategically using enzymes on meat as a substrate. The beef protein hydrolysates contain bioactive peptides with antihypertensive, antioxidant, and antibacterial properties. Transglutaminase (TGase) enzymes are used for improving the texture of a variety of food products (Singh et al. 2019). It is a calcium-dependent enzyme catalyzing acyl transfer process, comprising of the amino group lysine, with glutamine–carboxamide groups acting as acyl donors and lysyl residues acting as acyl acceptors. In dairy, baking, brewing, and wine making industries enzymes have been utilized for ages but not in isolated form. In order to manufacture cheese and other milk-related products, enzymes are necessary. Enzymes are also employed to reduce the amount of alcohol and calories in beer. In wine making, enzymes are used for reducing sulfur content, preservation of color and flavor, and clarification of wine (Spohner et al. 2015). Enzymes are added either singly or in complicated mixes at a very minimal level for the production of baked products. Lipolytic enzymes are increasingly used in the baking sector. On an industrial scale, a vast number of fat-clearing enzymatic lipases are manufactured. They are used mainly in industry for adding flavor to dairy products and processing of foods like vegetables, fruits, meat, milk products and beer. Recent discoveries imply that phospholipases can be utilized for enhancing or replacing standard emulsifiers since they breakdown wheat lipids for the formation of emulsified lipids (Meshram et al. 2019). In egg yolk treatment, iso-lecithin is produced by hydrolyzing lecithin using the enzyme phospholipase, thus enhancing the emulsification property and increasing heat stability. Custard, baby food, mayonnaise, and dough preparation can all benefit from the egg yolk made this way (Aravindan et al. 2007). Starch is the primary component in bread. The inclusion of amylase and lipase enzymes to bread production lowers the starch crystallinity in the bread, thus extending its shelf life. Glutaminase is essential (L-glutamine aminohydrolase) flavor-adding enzyme in meat sausage produced by starting cultures. Glutamic acid level increases when we add enzyme to fermented products, intensifying the "umami" flavor of the dish. Hydrolytic deamination of L-glutamine occurs by glutaminase resulting in the formation of L-glutamic acid. This is a flavor enhancer as well as an acid neutralizer. Although it is a common enzyme in bacteria and eukaryotes, its prevalence in thermophiles, archaea, and plants is unknown (Singh et al. 2019). Difructose anhydride (DFA) III is a disaccharide that is non-carcinogenic and non-digestible but helps the intestine in the absorption of calcium, magnesium, and other minerals. Inulin fructotransferase has since been discovered in *Arthrobacter* sp. and other bacteria. However, DFA III's industrial usage was hampered by its low thermal stability and the high cost of inulin. Fructooligosaccharides (FOS) are prepared from sucrose by using fructosyltransferase enzyme. The smallest substrate for inulin fructotransferase FOS is known to possess a fructofuranosidic linkage identical to inulin. As a unique technique, a linked enzyme reaction utilizing sucrose as substrate to synthesize DFAIII, yielding roughly 10% (w/w) (Choi et al. 2015). Papain is a very effective enzyme that degrades myofibrillar and collagen proteins significantly. It is reported that papain showed its best activity in pH levels and temperatures of around 5.8–7 and 50–57 °C respectively especially when casein was employed as the substrate (Singh et al. 2019). Ficin is a protease found in the latex of *Ficus anthelmintica, Ficus glabrata*, and other Ficus species having 1.5 h half-life at 60 °C. It is mostly derived from fig fruit that helps in the tenderization of meat. Actinidin or actinidain enzyme is found in kiwi fruit and its concentration differs depending on the different fruit varieties. It showed its best activity between the temperatures and pH ranging from at 58 to 62 °C and 7–10 respectively. Both connective tissue and myofibrillar proteins are known to be hydrolyzed using actinidin but proteolytic activity of collagen is much higher (Singh et al. 2019) (Table 2.1).

2.4 Production of Flavors by Enzymatic Technology

Food flavors are one of the most common types of food additives and are used for increasing the flavor of food products. Enzymes as biocatalysts open up a world of possibilities for food flavor creation. Enzymes can also be used directly as food additives, not just to synthesize or liberate flavor from precursors, but also to

Table 2.1 Enzymes, their sources, and applications in food industries

Source	Enzymes	Food applications	References
Aspergillus niger	Aminoacylase and aminopeptidase	(a) Protein hydrolysis (b) Enhancing flavor	Basso and Serban (2020)
Aspergillus oryzae, Bacillus subtilis, Aspergillus niger	α and β amylase	(a) Distilling, baking, and brewing industry—modification of dough and starch hydrolysis	Villas-Boas and Franco (2016)
Mortierella vinaceae	α-Galactosidase	(a) Helps in digestion of complex sugar and fat	Kote et al. (2020)
Aspergillus sp.	α-Glucosidase	Starch processing enzyme	Şöhretoğlu and Sari (2020)
Aspergillus sp. *Penicillium* sp.	β-Glucosidase	Lignocellulose degradation and also flavor enhancement in wines	Zhang et al. (2021)
Rhizopus oryzae	Carbohydrase	(a) Production of dextrose from starch (b) Clarification of fruit juices (c) Starch removal from wort (d) Viscosity control of chocolate syrup	Peña-Lucio et al. (2021)
Aspergillus niger, Micrococcus lysodeikticus	Catalase	(a) Removal of hydrogen peroxide from milk prior to cheese making	Raveendran et al. (2018)
Aspergillus niger, Aspergillus nidulans	Cellulase	(a) Liquid coffee concentrate (viscosity control) (b) Clarification and stabilization of juices and wines	Sharma et al. (2017)
Mucor miehei	Esterase	(a) Hydrolyze ester group into an acid and alcohol	Rodrigues and Fernandez-Lafuente (2010)
Aspergillus niger, Pseudomonas paucimobilis	Naringinase	(a) Bitter taste removal in fruit juice and enhancing aroma of wine	Yadav et al. (2021)
Aspergillus usamii	Protease	(a) Clarification of beer and controls the viscosity (b) Dough modification (c) Meat tenderization	Aruna et al. (2014)
Aspergillus niger	Pectinase	(a) Fruit juice and wine (production and clarification)	Shrestha et al. (2021)
Pancreatic tissue and edible forestomach of calves	Lipase	(a) Cheddar cheese production	Szymczak et al. (2021)
Aspergillus oryzae, Candida rugosa, Thermomyces lanuginosus		(b) Cheese flavor development	

(continued)

Table 2.1 (continued)

Source	Enzymes	Food applications	References
Animal—pancreas of pig	Trypsin	(a) Digestion of animal and milk proteins	Geisslitz et al. (2022)
Aspergillus fijiensis, *Saccharomyces* sp.	Invertase	(a) Sucrose hydrolysis	Manoochehri et al. (2020)
Aspergillus sp.	Tannase	(a) Use in tea products and to enhance antioxidant properties	Lekshmi et al. (2021)
Trichoderma sp.	Pullulanase and Rhamnosidase	(a) Used in starch (amylopectin) processing	Xu et al. (2021)
Penicillium sp.		(b) Used in flavor development (e.g., in wines) and debittering	
Leptographium	Phosphodiesterase	(a) Used as a flavor enhancer	Ahmed et al. (2021)
Penicillium			

eliminate off-flavors caused by naturally occurring or synthesized substances (Bigelis 1992). Their specificity, whether applied via whole-cell or cell-free systems, enables the production of certain compounds that are difficult to synthesize; their stereoselectivity is a major benefit for the food industry, where a unique optical conformation may be related with flavor qualities. Whitaker (1990) provides a broad overview of the future of enzymes in food technology in general. Enzymes involved in flavoring can be endogenous, meaning they are naturally present in the food, or they can come from microbial sources, either deliberately introduced to foods or as a result of contamination. Flavoring substances and preparations, smoke flavorings, flavoring adjuvants, and process flavorings are examples of flavor additives (Müller 2007). Flavor esters are the most commonly utilized in the food and biopharmaceutics industries. Alcohols along with various combinations of small-chain carboxylic acids form flavor esters through esterification processes, resulting in distinct flavor and fragrance qualities (Sá et al. 2017). Isoamyl, ethyl, geranyl, citronellyl, and thiohexyl esters are some of the flavor substances generated in solvent free enzymatically conditions. Table 2.2 summarizes the flavoring compounds like esters and their sources. Immobilized lipases are favored for synthesis of ester due to various advantages including increased reusability, catalytic reaction, and greater stability even during agitation and heat (Mathesh et al. 2016). Also, because of its sweet banana flavor, isoamyl acetate is widely used in the food sector and is found in a wide variety of foods, including beverages and honey (Novak et al. 2016). Hexyl butyrate which is a chemical compound used in food and biopharmaceutics industries having fragrance and aroma is basically produced by transesterifying tributyrin and hexanol using immobilized lipase enzyme from *Rhizomucor miehei* (Chang et al. 2003).

Table 2.2 Source of flavoring compounds and their characteristics

Source	Flavoring compound	Characteristic	References
Hog pancreas	Isoamyl esters	Banana fruity flavor	Jaiswal and Rathod (2022)
Candida antarctica	Citronellyl esters	Fruity flavor	Zhang et al. (2020)
		Rose-fruity flavor	
Candida antarctica	Butyl caprylate	Fruity flavor	Sousa et al. (2021)
Pseudomonas sp.	Ethyl hexanoate	Fruity pineapple flavor	Wang et al. (2018a, b)
Pseudomonas fluorescens	Cinnamyl acetate	Spicy type flavor slightly fruity	Wang et al. (2018a, b)

2.5 Rules and Regulations for Application of Additives in Food

The majority of the foods we eat have been processed in some way or the other. Generally, all foods contain food additives to increase the aroma and shelf life. They are essential for maintaining food quality and increasing the acceptance rate among consumers. According to Codex Alimentarius, food additives are not to be consumed as food but rather can only be utilized as food ingredients (Codex 2021; EFSA 2021; FDA 2004). WHO (World Health Organization) and FAO (Food and Agriculture Organization) initiated the JECFA, i.e., Joint FAO/WHO Expert Committee on Food Additives, to evaluate the safety criteria and daily allowance limit of additives in food. JECFA conducted various tests to study the safety of food additives in terms of their impact on human health, bioavailability, and bioaccessibility at a particular dose (FAO 2021). Therefore, it is mandatory to mention the food additives used in foods on the packaging labels. Food flavors, colors, preservatives, enzymes, sugar substitutes, and structural and texture enhancers are among them (Saltmarsh 2020).

2.6 Conclusion

This chapter has focused on enzymes and its application in dairy, beverage, and other food industries. It is generally used in the food industry as it has GRAS status. Enzymes are also used as food additives as they can be directly added to food products for enhancing aroma, flavor, and increasing shelf life. They are proteins that help in boosting the biochemical reactions by breaking down the complex and larger molecules into their simpler forms. The sources of enzymes can either be plants, animals, or microbes. As compared to plant and animal enzymes, enzymes from microbes can be produced efficiently using solid-state and submerged fermentation techniques. Production of microbial enzymes on a large scale is much easier than others. Various biological and molecular techniques can be used to modify microbial enzymes. Food manufacturers are continuously exploring newer and advanced technologies for the production of enzymes to meet the growing demand and market among the consumers. Flavors, colors, sweeteners, stabilizers, and antioxidants are natural additives that are synthesized using enzymes such as lipase, amylase, protease, and xylanases. As a result, enzymatic synthesis is still a green, sustainable and efficient way to make food additives.

References

Abada EA (2019) Application of microbial enzymes in the dairy industry. In: Enzymes in food biotechnology. Elsevier, Amsterdam, pp 61–72

Agyei D, Akanbi TO, Oey I (2019) Enzymes for use in functional foods. In: Enzymes in food biotechnology. Elsevier, Amsterdam, pp 129–147

Ahmed WS, Geethakumari AM, Biswas KH (2021) Phosphodiesterase 5 (PDE5): structure-function regulation and therapeutic applications of inhibitors. Biomed Pharmacother 134: 111128

Akanbi TO, Barrow CJ, Byrne N (2012) Increased hydrolysis by *Thermomyces lanuginosus* lipase for omega-3 fatty acids in the presence of a protic ionic liquid. Catal Sci Technol 2:1839–1841

Akanbi TO, Ji D, Agyei D (2020) Revisiting the scope and applications of food enzymes from extremophiles. J Food Biochem 44:13475

Aravindan R, Anbumathi P, Viruthagiri T (2007) Lipase applications in food industry. Indian J Biotechnol 6:141–158

Aruna K, Shah J, Birmole R (2014) Production and partial characterization of alkaline protease from *Bacillus tequilensis* strains CSGAB 0139 isolated from spoilt cottage cheese. Int J Appl Biol Pharm 5:201–221

Basso, A., & Serban, S. (2020) Overview of immobilized enzymes' applications in pharmaceutical, chemical, and food industry. Immobilization of enzymes and cells, 27-63

Bhavsar KV, Yadav GD (2019) Synthesis of geranyl acetate by transesterification of geraniol with ethyl acetate over *Candida antarctica* lipase as catalyst in solvent-free system. Flavour Fragrance J 34:288–293

Bigelis R (1992) Flavor metabolites and enzymes from filamentous fungi. Food Technol 46(10): 151–161

Chang S-W, Shaw J-F, Shieh C-J (2003) Optimization of enzymatically prepared hexyl butyrate by lipozyme IM-77. Food Technol Biotechnol 41:237–242

Choi J-M, Han S-S, Kim H-S (2015) Industrial applications of enzyme biocatalysis: current status and future aspects. Biotechnol Adv 33:1443–1454

Codex (2021). http://www.codexalimentarius.org/

EFSA (2021). https://www.efsa.europa.eu/sites/default/files/2021-12/7033.pdf

FAO (2021). https://www.fao.org/publications/sofa/sofa-2021/en/

FDA (2004). https://www.fda.gov/food/food-allergensgluten-free-guidance-documents-regulatory-information/foodallergen-labeling-and-consumer-protection-act-2004-questions-and-answers

FDA (2020). https://www.fda.gov/industry/color-additive-inventories/color-additive-status-list

Geisslitz S, Weegels P, Shewry P, Zevallos V, Masci S, Sorrells M, Brouns F (2022) Wheat amylase/trypsin inhibitors (ATIs): occurrence, function and health aspects. Eur J Nutr:1–8

Gurung N, Ray S, Bose S, Rai V (2013) A broader view: microbial enzymes and their relevance in industries, medicine, and beyond. BioMed Research International, 2013

Homaei A (2016) Enzyme immobilization and its application in the food industry. In: Advances in food biotechnology, 1st edn. Wiley, Hoboken, pp 145–164

Huang SM, Huang HY, Chen YM, Kuo CH, Shieh CJ (2020) Continuous production of 2-phenylethyl acetate in a solvent-free system using a packed-bed reactor with Novozym 435. Catalysts 10:714

IHS Markit (2020). https://ihsmarkit.com/products/chemical-food-additives-scup.html

Jaiswal KS, Rathod VK (2022) Process intensification of enzymatic synthesis of flavor esters: a review. Chem Rec 22(3):e202100213

Katrolia P, Rajashekhara E, Yan Q, Jiang Z (2014) Biotechnological potential of microbial α-galactosidases. Crit Rev Biotechnol 34(4):307–317

Kote N, Manjula AC, Vishwanatha T, Patil AGG (2020) High-yield production and biochemical characterization of α-galactosidase produced from locally isolated Penicillium sp. Bull Nat Res Centre 44(1):1–10

Lekshmi R, Nisha SA, Vasan PT, Kaleeswaran B (2021) A comprehensive review on tannase: microbes associated production of tannase exploiting tannin rich agro-industrial wastes with special reference to its potential environmental and industrial applications. Environ Res 201: 111625

Li Z, Scott K, Hemar Y, Zhang H, Otter D (2018) Purification and characterisation of a protease (tamarillin) from tamarillo fruit. Food Chem 256:228–234

Lin X, Ma L, Racette SB, Anderson Spearie CL, Ostlund RE Jr (2009) Phytosterol glycosides reduce cholesterol absorption in humans. Am J Physiol Gastrointest Liver Physiol 296:G931–G935

Manoochehri H, Hosseini NF, Saidijam M, Taheri M, Rezaee H, Nouri F (2020) A review on invertase: its potentials and applications. Biocatal Agric Biotechnol 25:101599

Martins SL, Albuquerque BF, Nunes MAP, Ribeiro MHL (2018) Exploring magnetic and imprinted cross-linked enzyme aggregates of rhamnopyranosidase in microbioreactors. Bioresour Technol 249:704–712

Mathesh M, Luan B, Akanbi TO, Weber JK, Liu J, Barrow CJ, Zhou R, Yang W (2016) Opening lids: modulation of lipase immobilization by graphene oxides. ACS Catal 6:4760–4768

Meshram A, Singhal G, Bhagyawant SS, Srivastava N (2019) Plant-derived enzymes: a treasure for food biotechnology. In: Enzymes in food biotechnology. Elsevier, Amsterdam, pp 483–502

Müller DA (2007) Flavours: the legal framework, flavours and fragrances. Springer, Berlin, pp 15–24

Novak U, Lavric D, Žnidaršiè-Plazl P (2016) Continuous lipase B-catalyzed isoamyl acetate synthesis in a two-liquid phase system using corning AFR module coupled with a membrane separator enabling biocatalyst recycle. J Flow Chem 6:33–38

Okolie CL, Akanbi TO, Mason B, Udenigwe CC, Aryee AN (2019) Influence of conventional and recent extraction technologies on physico-chemical properties of bioactive macromolecules from natural sources: a review. Food Res Int 116:827–839

Pandey A, Nigam P, Soccol CR, Singh D, Soccol VT, Mohan R (2000) Advances in microbial amylases. Biotechnol Appl Biochem 31:135–152

Patel AK, Singhani RR, Pandey A (2016) Novel enzymatic processes applied to the food industry. Curr Opin Food Sci 7:64–72

Peña-Lucio E, Chávez-González ML, Londoño-Hernandez L, Govea-Salas M, Abdulhameed S, Aguilar CN (2021) Microbial caffeinases: biosynthesis and properties. In: Microbial and natural macromolecules. Academic Press, pp 87–99

Ramadan MF (2019) Enzymes in fruit juice processing. In: Enzymes in food biotechnology. Elsevier, Amsterdam, pp 45–59

Raveendran S, Parameswaran B, Ummalyma SB, Abraham A, Mathew AK, Madhavan A, Pandey A (2018) Applications of microbial enzymes in food industry. Food Technol Biotechnol 56(1): 16

Rodrigues RC, Fernandez-Lafuente R (2010) Lipase from *Rhizomucor miehei* as an industrial biocatalyst in chemical process. J Mol Catal B Enzym 64(1–2):1–22

Saltmarsh M (2020) Food additives and why they are used, pp 1–9

Sá AGA, de Meneses AC, de Oliveira D, de Araújo PHH (2017) A review on enzymatic synthesis of aromatic esters used as flavor ingredients for food, cosmetics and pharmaceuticals industries. Trends Food Sci Technol 69:95–105

Sharma S, Kuila A, Sharma V (2017) Enzymatic hydrolysis of thermochemically pretreated biomass using a mixture of cellulolytic enzymes produced from different fungal sources. Clean Techn Environ Policy 19(5):1577–1584

Shrestha S, Rahman M, Qin W (2021) New insights in pectinase production development and industrial applications. Appl Microbiol Biotechnol 105(24):9069–9087

Singh PK, Shrivastava N, Ojha BK (2019) Enzymes in the meat industry. In: Enzymes in food biotechnology. Elsevier, Amsterdam, pp 111–128

Şöhretoğlu D, Sari S (2020) Flavonoids as alpha-glucosidase inhibitors: mechanistic approaches merged with enzyme kinetics and molecular modelling. Phytochem Rev 19(5):1081–1092

Sousa RR, Silva AS, Fernandez-Lafuente R, Ferreira-Leitao VS (2021) Solvent-free esterifications mediated by immobilized lipases: a review from thermodynamic and kinetic perspectives. Cat Sci Technol 11:5696–5711

Spohner SC, Müller H, Quitmann H, Czermak P (2015) Expression of enzymes for the usage in food and feed industry with Pichia pastoris. J Biotechnol 202:118–134

Sun L, Xin F, Alper HS (2021) Bio-synthesis of food additives and colorants—a growing trend in future food. Biotechnol Adv 47:107694

Szymczak T, Cybulska J, Podleśny M, Frąc M (2021) Various perspectives on microbial lipase production using agri-food waste and renewable products. Agriculture 11(6):540

Trani A, Loizzo P, Cassone A, Faccia M (2017) Enzymes applications for the dairy industry. In: Industrial enzyme applications. Wiley, Hoboken, pp 166–175

Villas-Boas F, Franco CM (2016) Effect of bacterial β-amylase and fungal α-amylase on the digestibility and structural characteristics of potato and arrowroot starches. Food Hydrocoll 52:795–803

Viniegra-González G, Favela-Torres E, Aguilar CN, Romero-Gomez S, Díaz-Godinez G, Augur C (2003) Advantages of fungal enzyme production in solid state over liquid systems. Biochem Eng J 13:157–167

Wang X, Wang X, Cong F, Xu Y, Kang J, Zhang Y, Zhou M, Xing K, Zhang G, Pan H (2018a) Synthesis of cinnamyl acetate catalysed by highly reusable cotton-immobilized Pseudomonas fluorescens lipase. Biocatal Biotransformation 36:332–339

Wang Y, He L, Pan Q, Duan C, Wang J (2018b) Effects of basal defoliation on wine aromas: a meta-analysis. Molecules 23(4):779

Whitaker JR (1990) New and future uses of enzymes in food processing. Food Biotechnol 4(2): 669–698

Xu P, Zhang SY, Luo ZG, Zong MH, Li XX, Lou WY (2021) Biotechnology and bioengineering of pullulanase: state of the art and perspectives. World J Microbiol Biotechnol 37(3):1–10

Yadav M, Sehrawat N, Sharma AK, Kumar V, Kumar A (2021) Naringinase: microbial sources, production and applications in food processing industry. J Microbiol Biotechnol Food Sci:717–720

Zhang Y, He S, Simpson BK (2018) Enzymes in food bioprocessing—novel food enzymes, applications, and related techniques. Curr Opin Food Sci 19:30–35

Zhang S, Guo F, Yan W, Dong W, Zhou J, Zhang W, Jiang M (2020) Perspectives for the microbial production of ethyl acetate. Appl Microbiol Biotechnol 104(17):7239–7245

Zhang P, Zhang R, Sirisena S, Gan R, Fang Z (2021) Beta-glucosidase activity of wine yeasts and its impacts on wine volatiles and phenolics: a mini-review. Food Microbiol 100:103859

Chapter 3
Microbial Antioxidants in Food Products

Diksha Sood, Sunita Devi, Bindu Devi, and Priyanka Arya

Abstract Microbial metabolites have already established themselves as the lifesaving medications on account of their usage in curing many diseases. Microbial antioxidants are naturally synthesised by microbes that could be exploited to assist humans by scavenging harmful free radicals from the body. Free radicals (FRs) have a great degree of reactivity, forming interactions with biological constituents like DNA and lipids, causing tissue damage. Once free radical concentration upsurges, oxidation stress conditions prevail in the host body which are accountable for various diseases. However, antioxidants assist in maintaining a balance between antioxidants and FRs for eluding oxidative stress. Antioxidants are mainly enzymatic or non-enzymatic. A plenitude of microbes viz. Bacteria (*Lactobacillus* spp., *Pseudomonas hibiscicola, Enterobacter ludwigii, Macrococcus caseolyticus, Bacillus anthracis*); actinomycetes (*Streptomyces prunicolor, S. melanogenes, S. prunicolor*); Blue green algae (*Nostoc*, *Anabaena*, *Arthrospira, Haematococcus pluvialis*); fungi (*Colletotrichum* spp., *Mortierella* spp., *Penicillium* spp., *Eurotium* spp., *Cladosporium* sp., *Aspergillus* spp.); and yeasts (*Candida utilis, Kluyveromyces marxianus, Yarrowia lipolytica, Pichia kudriavzevii*) exhibit the potency to produce antioxidants naturally. This chapter highlights the impending usage of microbes to produce antioxidants as metabolites plus their possibility to embody a splendid source of natural antioxidants.

Keywords Antioxidants · Free radicals · Oxidative stress · Microbial antioxidants · Microbial antioxidants in food products

D. Sood · S. Devi (✉) · B. Devi
Department of Basic Sciences, College of Forestry, Dr. YSP University of Horticulture & Forestry Nauni, Solan, Himachal Pradesh, India

P. Arya
Department of Food Science and Technology, College of Horticulture, Dr. YSP University of Horticulture & Forestry Nauni, Solan, Himachal Pradesh, India

A. K. Nadda, G. Goel (eds.), *Microbes for Natural Food Additives*, Microorganisms for Sustainability 38, https://doi.org/10.1007/978-981-19-5711-6_3

3.1 Introduction

Antioxidants are chemical elements that exist in the animal's body and assist in lessening the destruction triggered via free radicals of oxygen (O) and nitrogen (N) produced in the host body due to various metabolic processes (Bensid et al. 2022). Although the occurrence of oxygen is an ultimate aspect for life's manifestation on the Earth, its presence in the host body as free radicals (FRs) may be detrimental as it may cause massive health problems (Chandra et al. 2020). Oxidative stress (OS) is described as a disruption in the pro-oxidants-antioxidants encompasses systems that favours the former (Sies 1985; Rani et al. 2021). To put it another way, oxidative stress (OS) is recognised as the state where free radicals quantity upsurges antioxidants quantity in the humans body (Kumar et al. 2011; Singh et al. 2014) that causes harm to body's cells plus tissues on account of their ability to bond effortlessly with different metabolites of processes happening inside body besides deviation from the end product. Antioxidants also behave like immunity booster besides preventing cancer and other innumerable diseases (Amany and Abdel-Raheam 2020).

Antioxidants primarily use two types of defence mechanisms, viz. non-enzymatic and enzymatic to hunt down FRs. Enzymatic mechanism involves the employment of SOD (superoxide dismutase), GSH peroxidase, and catalase enzymes (Moussa et al. 2019), whereas the employment of vitamin A, lycopenes, vitamin C, β-carotene, vitamin E plus other molecules represents a non-enzymatic defence mechanism (Kumar et al. 2018). Antioxidants simply disrupt or inhibit the chain reaction that yields FRs, preventing OS via giving or accepting electrons, metal chelation of FRs as soon as they are created in the host's body through metabolic processes (Barbalho et al. 2022). Fruits plus vegetables contain plentiful amounts of naturally occurring antioxidants, which supply a great extent of the antioxidants required by the host body; hence, they must be included in one's diet to reap health benefits (Sdona et al. 2022). People who consume antioxidants-rich foods as their usual diet are less vulnerable to enduring diseases (Arias et al. 2022; Dembinska-Kiec et al. 2008) and live longer lives than those who do not consume antioxidants in their diets. Fortunately, since our regular diet includes an adequate amount of fruits and vegetables, the antioxidants present in them provide us with several health benefits, such as protection against diverse harmful and chronic diseases (Sdona et al. 2022).

Vitamins A and C aid in the slowing of ageing, the deterrence of cataracts, and reducing the risk of respiratory and cardiovascular disorders, including chronic heart attack, ischaemic and cardiac stroke besides shielding tissues and cells from free radical-mediated damage during OS (Timoshnikov et al. 2022; Shati et al. 2021). A huge number of supplements are certainly available on the market, but increased antioxidant levels have a detrimental impact on consumer health; therefore we should only incorporate these supplements into our regular diets after consulting with a dietician or physician (Jîtca et al. 2022). Antioxidants are very stable food components found in raw food that also work as a preserving agent in fresh fruits plus vegetables straight from the farm to our plates. Antioxidants are indispensable for the endurance of both plants and animals, therefore making them a part in our diets so that these can assist us in improving our well-being health wise (Sdona et al. 2022). Examples of some food items that contain great amounts of antioxidants are wild and cultivated blueberries, beans (Pinto, small red, black, and red kidney), cranberries, artichokes, prunes, blackberries, strawberries, raspberries, Granny Smith, red delicious, and Gala apples, tomatoes, pecans, black plums, sweet cherries, Russet potatoes, and leafy greens (Arias et al. 2022; Bensid et al. 2022).

3.2 Free Radicals

Recently, the emphasis of scientists has switched to occurrence of free radicals in humans, because they are answerable to the aetiology of many health-related disorders and can add new aspects to disease management. Atoms/molecules with a free or unpaired electrons (e^-) in the outermost atomic orbital or valence shell are denoted as FRs. Since these unpaired electron's nature is exceedingly reactive and unstable, they must either obtain an e^- from another species or offer their e^- to other species to achieve stability. Free O and N radicals are acknowledged as reactive O species (ROS) and reactive N species (RNS), respectively (Han et al. 2018).

Both endogenous and exogenous sources contribute towards RNS and ROS production. Examples of exogenous sources include industrial solvents, pollution, tobacco smoke, pesticides, heavy metals, etc. whereas endogenous sources include endoplasmic reticulum, mitochondria, and phagocytic cells besides metabolic pathways occurring in these cellular organelles that release several FRs inside the host's body. Furthermore, enzymatic sources of FRs are phagocytosis process, respiratory chain, cytochrome P_{450} system, and synthesis of prostaglandins. Therefore, FRs are primarily their end result while also being a component of ordinary cellular metabolism. Figure 3.1 portrays the endogenous and exogenous sources of FRs (Li et al. 2021).

These FRs are extremely reactive species, possessing the capability of damaging nucleus and cell's membranes besides cell components like lipids, DNA, proteins,

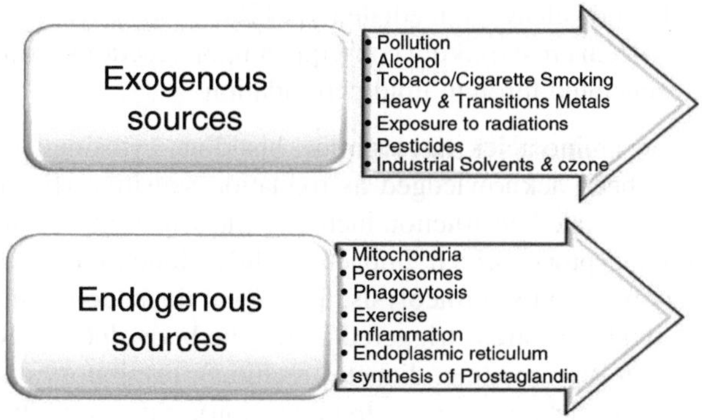

Fig. 3.1 Exogenous and endogenous sources of FRs

and carbohydrates (Gulcin 2020). The commonly present ROS inside body are hydroxyl (OH), alkoxyl (RO), superoxide (O_2^-), hydrogen peroxide, peroxyl radical, hypochlorous acid, organic peroxide, and Ozone (O_3) while RNS are nitric oxide (NO), nitrous acid, nitrogen dioxide, nitrosyl cation/anion (NO^+/NO^-), nitryl chloride plus peroxynitrite (ONOO).

FRs contribute significantly towards host defence system in eradicating those pathogens which exist in reasonable or low concentrations. Free radicals produced during cellular processes aid in pathogen phagocytosis and hence aid the immune system's defence line in preventing harmful pathogens from causing disease. They are thus an important constituent of the immunogenic system. FRs are also identified for their dynamic function in cell signalling process. The NO is the usual cell signalling molecule used for cell to cell signalling for various processes like thrombosis, neural activity, and blood flow mechanism (Parker et al. 2007). Nitric oxide also assists in cancer cell's removal from the body. Another significant function of FRs includes granulomatous disease. Since individuals with this condition are inept to form $O_2^{\bullet-}$ molecule because of a malfunctioning NADPH oxidase system, this makes them prone to an array of diseases. Thus, free radical's presence in the host system in reasonable concentrations is advantageous to humans by preventing an assortment of ailments.

3.3 Oxidative Stress

OS symbolises that state of host body when a discrepancy befalls in the FRs' utilisation and production or little concentrations of antioxidants inside body. We have been conscious of the statistics that oxidative DNA plays a main function in causing cancer. However, interim OS could be caused by infection, trauma, hyperoxia, heat injury, excessive exercise, and toxins. Augmented levels of RNS and ROS showed harmful impacts on cell's structures, including proteins, nucleic acids, and lipids (Wu et al. 2013). The damage caused to the structures of proteins/nucleic acids (DNA plus RNA) is mainly categorised into three ways:

1. Alteration of a particular amino acid via oxidation
2. Peptide cleavage mediated via FRs
3. Protein cross-linking consequent upon reactions with metabolic reaction products generated through lipid peroxidation

A few amino acids viz. arginine, histidine, cytosine, and methionine of the proteins have been acknowledged as oxidation sensitive (Freeman and Crapo 1982). The FRs-mediated destruction increases the sensitivity of enzymes, membrane transport, and receptors, which impacts cellular functioning. Variations in processes like enzyme activity, proteolysis, and signal transduction mechanisms lead to ageing.

Oxidative stress triggered inside the body due to FRs is acknowledged to cause various diseases (e.g. diabetes mellitus); respiratory diseases (e.g. asthma); cardiovascular sicknesses (e.g. hypertension); cancerous diseases (e.g. breast, prostate,

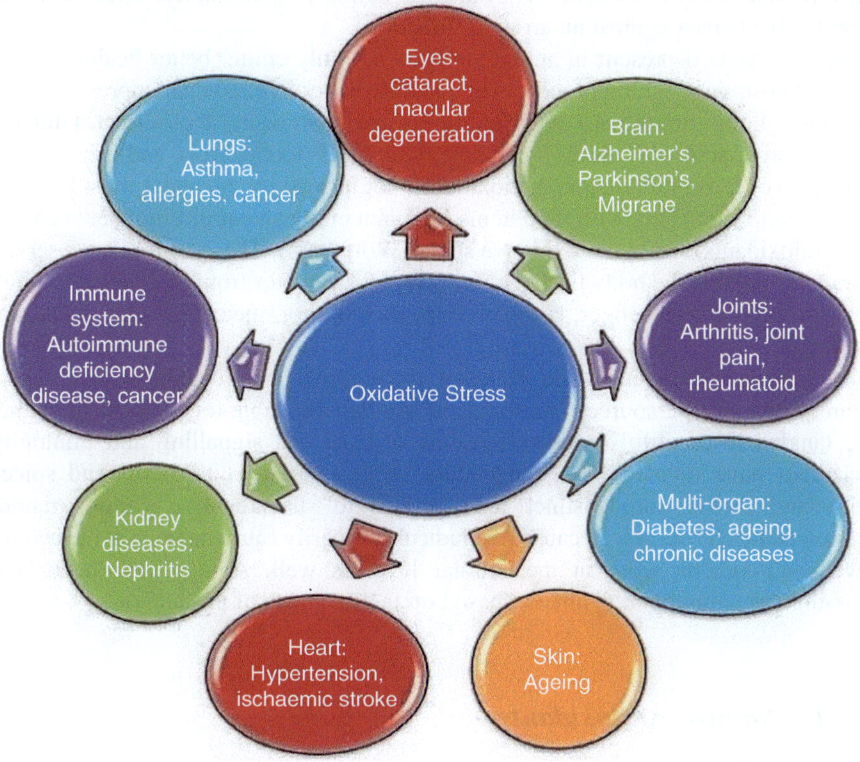

Fig. 3.2 Oxidative stress-mediated diseases

and, bladder cancer); neurodegenerative disorders (e.g. multiple sclerosis, Parkinson and Alzheimer diseases); besides cataract, atherosclerosis, and rheumatoid arthritis (Fig. 3.2). It also performs mitochondrial DNA's oxidation, which is indirectly related to an array of health disorders, including cancer. The manifestation of FRs is obligatory for lipid peroxidation to continue, but under OS conditions, their secondary sources are present, and the reaction chain continues, ensuing the creation of a great extent of secondary metabolic products like alkanes, aldehydes, and isoprostanes.

3.4 Antioxidants

These are chemical elements that exist in the host body and assist in scavenging or neutralising FRs and curing/combating OS conditions. With people's growing concern for their health and the path to living a healthy life, antioxidants have received increasing attention. Including antioxidants in their regular diets can protect them from an array of ailments. They also aid in boosting immunity, which works as a protective barrier against an array of infections.

Antioxidants if present in ample amounts not only confer better health benefits but also postpones ageing process. They mainly lower free radicals impression from the host's body through scavenging process and neutralising their effect. Glutathione, ascorbic acid, ubiquinol, β-carotene, vitamin E (tocopherol), and uric acid are examples of few antioxidants. Antioxidants, not formed by the body, must be a part of a person's diet along with food items for maintaining an equilibrium between FRs and antioxidants, and avoid oxidative stress (Wolf 2005). These antioxidants serve a variety of tasks in the body like electron giver and acceptor from free radicals, single oxygen molecule scavenger, H-donor, enzyme inhibitor, metal chelating agent, and synergist.

Antioxidants can easily be obtained from fruits and vegetables like natural sources or synthetic sources. Antioxidants are indispensable for the host since they are tangled in plentiful complex systems such as cell signalling and immunity. Scientists have divided fruits, vegetables, drinks, dried fruits, herbs and spices, nuts, and cereals into distinct food groups to subcategorize various natural antioxidant-rich sources. Because free radicals primarily cause damage at the cellular level, antioxidants work at the cellular level as well. Antioxidants have been classified as natural and synthetic, based on their sources of production.

3.4.1 Natural Antioxidants

Antioxidants that emanate from natural sources are known as natural antioxidants. They are primarily phenol-derivative compounds that exist in the whole plant parts, including nuts, fruits, seeds, roots, vegetables, and barks. Vitamin E (fat soluble vitamin) assists in fat's oxidation to provide health advantages to humans. The α-tocopherol is a synthetic vitamin E supplement that assists in absorbing vitamin E from natural sources. Vitamin E aids in eluding the membranes' lipid peroxidation. Its primary role is to scavenge the peroxy radical during lipid peroxidation.

Citrus fruits are the richest sources of vitamin C, which helps in reducing an enormous variety of water soluble molecules. Vitamin C is also known as ascorbic acid. Apart from the rejuvenation of tocopherol molecule, ascorbic acid lessens the tocopherol FRs as prime function. The enzyme, L-gulono-g-lactone oxidase, is engrossed in ascorbic acid's synthesis, and the persons who are lacking in this enzyme must supplement their diet with synthetic supplements. β-carotene, a retinol's precursor, is another antioxidant included in our diets. This vitamin A helps in excluding the oxidant molecules (Young and Lowe 2018).

3.4.2 *Synthetic Antioxidants*

Synthetic antioxidants are chemically synthesised compounds/molecules, which are incorporated for preventing oxidation reactions in the food products like lipid peroxidation. Butyl-1-4-hydroxytoluene (BHT), propylgallate (PG), butylated hydroxyanisole (BHA), and *tert*-butyl hydroquinone (TBHQ) are all major antioxidants utilised in the food business, with TBHQ being the utmost proficient antioxidant (Pimpa et al. 2009). Both BHT and BHA are accountable for inhibiting chain reactions via free radical's stabilisation. The US, New Zealand, Australia, and the EU along with several other countries have authorised BHT and BHA as food antioxidants (García-García and Searle 2016). However, in a few nations, their usage has been approved within permissible limits. The employment of TBHQ is documented in animal fats and vegetable oils. Propyl galate is a synthetic antioxidant that is lipid soluble, gives oils a distinct aroma, and is heat sensitive. Table 3.1 depicts the antioxidant's sources (Sasse et al. 2009; Anwar et al. 2018).

Reliant upon the nature and action, antioxidants may be enzymatic or non-enzymatic. The latter type is further divided into two categories—nutrient and phytochemical antioxidants (Fig. 3.3), each of which is described as follows.

3.4.2.1 Enzymes as Antioxidants

These are endogenous antioxidants that help in removing FRs from the host body. The three primary enzymes entailed in removing antioxidants are superoxide dismutase (SOD), glutathione peroxidase (GSH), and catalase.

Table 3.1 Sources of antioxidants

Natural antioxidants		Synthetic antioxidants	
Leeks	Nuts	Octyl gallate	Ethoxyquin
Red wine	Leafy greens	Dodecyl gallate	Trihydroxybutyrophenine (THBP)
Green tea	Apricots	BHA (butylated hydroxyanisole)	PG (propyl gallate)
Citrus fruits	Carrots	BHT (butylated hydroxytoluene)	TBHQ (*tert*-butylhydroxyquinone)
Tomatoes	Pecans	Dilaurylthiodipropionate	Erythorbic acid (D-ascorbic acid)
Black plums	Sweet cherries	Thiodipropionic acid	Ascorbyl palmitate
Strawberries	Red kidney beans	2-Naphthol (2NL)	4-Phenylphenol

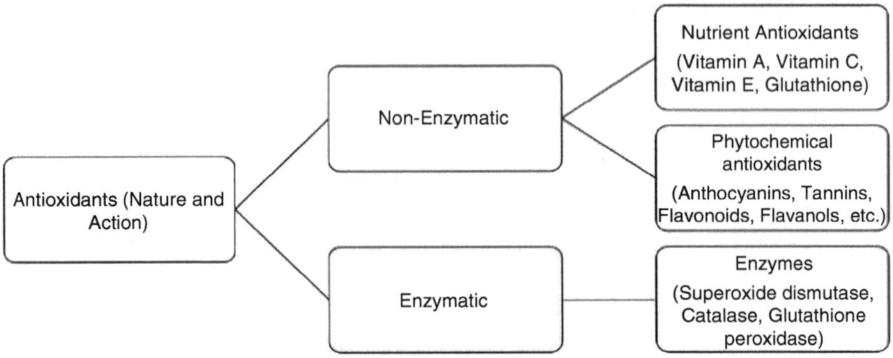

Fig. 3.3 Classification of antioxidants reliant on their nature and action

Superoxide Dismutase

The SOD enzyme plays an indispensable task in the body's defence mechanism by eliminating superoxide FRs and converting them to H_2O_2 (hydrogen peroxide) and oxygen (Boveris et al. 2006) as depicted in the reaction. The central sources of SOD are leafy greens, barley, broccoli, wheat, cabbage, etc. Superoxide dismutase exists in cytosol of eukaryotes, containing zinc plus copper as cofactor, whereas in mitochondria and bacteria, cofactor of SOD enzyme is manganese. In bacteria and cyanobacteria, iron is the cofactor of the SOD. As the extent of O_2 (oxygen) or free radicals increases, the SOD activity also increases.

$$2{O_2}^- + 2H^+ \overset{SOD}{\rightarrow} H_2O_2\ (\text{Hydrogen peroxide}) + O_2(\text{Oxygen})$$

Catalase Enzyme

The catalase is another key enzyme intricate in antioxidant action with utmost activity in liver, kidney, and connective tissues. It serves as FR scavengers, converting H_2O_2 to oxygen and water molecules as depicted in the reaction. This enzyme exists in all bodily organs and its activity increases as H_2O_2 levels rise (Gassen and Youdim 1999).

$$2H_2O_2 \overset{\text{Catalase}}{\rightarrow} H_2O + O_2$$

Glutathione Peroxidase

The third important enzyme linked to antioxidant mechanism is glutathione peroxidase. This enzyme aids in the exclusion of peroxidase radicals from the body because the peroxisomes are harmful FRs that attack lipids and DNA molecules. The glutathione is a flavoprotein that catalases the reducing reaction of glutathione disulphide (GSSG) using nicotinamide adenine dinucleotide hydrogen phosphate (NADPH) as revealed in the reaction. This enzyme has selenium as its cofactor that assists in eliminating peroxidase radicals from the host body (Ondrej et al. 2020).

$$\text{Free radicals}\begin{pmatrix} H_2O_2 \\ H_2O \end{pmatrix}\text{SOD/Se GPx}\begin{pmatrix} \text{GSSG} \\ \text{GSH} \end{pmatrix}\text{GSH Reductase P2}\begin{pmatrix} \text{NADP}^+ \\ \text{NADH} \end{pmatrix}\text{P3}$$

3.4.2.2 Nutrients as Antioxidants

Nutrient antioxidants are basically derived from antioxidant's exogenous sources which mainly include vitamins A, E, C, and glutathione.

Vitamin A

Vitamin A acts as both antioxidant and fat soluble antioxidant. β-carotene performs the function a predecessor for retinol (vitamin A) synthesis. This entails that β-carotene acts as a predecessor for retinol (vitamin A) synthesis. A synthesis can only occur if the essential precursor, β-carotene, is available. β-carotene occurs dominantly in an assortment of foods, viz. tomato, carrot, ash squash, peaches, yams, milk, butter, and yoghurt. Vitamin A inhibits oxidation reactions, and β-carotene assists in detoxifying the carcinogens accumulated in the liver besides lowering cancer's peril in the host (Solomons 2001).

Vitamin C

It is an aqueous solvable antioxidant that removes diverse types of FRs to escape oxidative stress situation. It is primarily responsible for converting oxidised free radicals to a reducing state. The vital enzyme required for its synthesis, L-gluno-γ-lactone oxidase, is lacking in our metabolism; hence ascorbic acid is not synthesised in human beings. Therefore, this must be encompassed in diet in surplus amount to gain their benefits as antioxidants (Singh et al. 2013).

Citrus fruits represent the rich store house of vitamin C. Other sources are fruit juices, kiwi, broccoli, cabbage, spinach, green pepper, strawberries, and kale. Since the quantity of vitamin C in extracellular matrix is higher, it plays a momentous role in extracellular protection. Ascorbic acid activity transforms the superoxide free radical form into dehydroascorbic acid, which is subsequently converted back to its indigenous form of vitamin C with the help of glutathione enzyme (Chambial et al. 2013).

Vitamin E

This vitamin has a dynamic function in protecting cellular or organelle's membranes from lipid peroxidation. Being lipid solvable antioxidant, it removes peroxy free radicals from the body synthesised during lipid peroxidation (Glenville et al. 2006). The principal vitamin E/tocopherol sources are nuts, seeds, vegetables, fish and fish oils, apricot etc.

Glutathione

It is a fundamental antioxidant for the defence mechanism against ROS existed in the tissues. Glutathione is a tripeptide molecule comprised of glutamyl, cysteinyl, and glycine subunits (Singh et al. 2013). In the tissues, it exists in an oxidised form that is reduced by glutathione reductase.

3.4.2.3 Phytochemicals as Antioxidants

Plants primarily create phytochemicals as a part of a defence mechanism against microbes and to lessen the OS conditions in plants that arise due to the existence of a massive number of FRs (Shukla et al. 2009). The phytochemical's involvement in lessening an extensive range of illnesses triggered by free radical's presence has extensively been recognised. On account of their redox potential, many phytochemicals have antioxidant capabilities, which help in reducing, absorbing, and breaking down peroxides and other free radicals inside the body. Lycopene, carotenoids, isoflavones, and flavonoids are certain well-known names of phytochemicals that serve as virtuous antioxidant's sources.

Anthocyanins

These are the pigments that have a violet, blue, or purple colour and are typically found in flowers, berries, and fruits. Petunidin, cyanidin, malvidin, delphinidin, peonidin, and pelargonidin represent the utmost prevalent dietary anthocyanins (Manach et al. 2004). Anthocyanins plus their derivatives could scavenge FRs via diverse mechanisms to reduce OS. Red cabbage anthocyanins, for instance, are affirmed for imparting protection from OS instigated by paraquat toxin (Igarashi and Kashiwagi 2000). In in vivo studies, cyanidin that is found in abundance in most fruits has shown to exhibit antioxidant activity (Tsuda et al. 2000). Another distinct study conducted on animals revealed that cyanidins provide protection to cell membrane's lipids against oxidation, triggered by numerous hazardous chemicals (Tsuda et al. 1999).

Flavonoids

These are the utmost prevalent secondary metabolites produced by higher plants, which could directly scavenge superoxide ions, OH radicals, and H_2O_2. Plants contain around 4000 different phenolic compounds naturally (Singh et al. 2017).

Flavanols

These represent a diverse category of polyphenols, which include monomeric flavan-3-ols plus polymeric procyanidins, often known as condensed tannins (dePascual-Teresa et al. 2010). Examples of most prevalent flavonols include Kaempferol, myricetin, laricitrin, isorhamnetin, and syringetin. Apple, lettuce, onion, broccoli, tomato, tea, grape, kale, red berries, and wine are all virtuous suppliers of flavonols. Greener leaves embrace a higher volume of flavonols (Manach et al. 2004). Flavonols confer numerous biological advantages to the host. For instance, they lower the risk of heart illness and cancer and improve endothelial function, besides lowering down the platelet activity. The antioxidant capabilities of all these plants are principally accountable for these feature (Patel 2008). Flavonols also aid in imparting protection to the lipids, cells, and DNA from oxidative destruction. The occurrence of aromatic rings in flavonoid molecule enables the release and uptake of e^- from FR species, imparting flavonols with antioxidant competencies.

Tannins

Tannins exist in an assorted kinds of foods, including fruits (e.g. blueberries, grapes, and persimmons) as well as tea, cocoa, legume fodders, and trees (e.g. *Acacia* spp., *Sesbania* spp.) besides grasses like sorghum and maize. Examples of tannins include Ellagitannins, Proanthocyanidins, and gallotannins. Proanthocyanidins discharge anthocyanidins, which contain antioxidant characteristics, when heated in alcohol or concentrated mineral acid. Tannins that could be hydrolysed include gallotannins and ellagitannins. Gallotannins are galloyl esters of quinic acid plus glucose, while ellagitannins are hexahydroxydiphenic acid derivatives (HHDP). The phloroglucinol, which is made up of phlorotannin subunits and found only in marine brown algae, denotes another example of tannin. Tannins impart a bitter/astringent flavour to food stuffs and beverages like unripe fruits, teas, and some red wines. Tannin's key function is not to supply hydrogen atoms or e^-, but to perform the task of secondary antioxidants via interfering with chain reactions or through chelation of iron like metal ions.

Phenolic Acids

These are basically a category of phenolic chemicals that occur in plants in abundant amounts. The most eminent phenolic acids are hydroxybenzoic acids (viz. *p*-hydroxybenzoic acid, protocatechuic acid, gallic acid, vanillic acid, and syringic acid) and hydroxycinnamic acids (viz. ferulic acid, chlorogenic acid, *p*-coumaric acid, sinapic acid, and caffeic acid) (Wrigstedt et al. 2010). Many therapeutic herbs, fruits, dietary spices, caffeic, vegetables, and cereals are acknowledged to contain ferulic plus *p*-coumaric acid. Ferulic acids are also abundant in wheat bran. These chemicals are utilised by plants in both structural and chemical defence measures antagonistic to OS (Cartea et al. 2011).

Due to enhanced opportunities for phenoxy radical delocalization, natural hydroxycinnamic acids possess higher quantity of antioxidants than hydroxybenzoic acid (Beer et al. 2002). The phenolic chemicals perform antioxidant's task in the human's biological systems through scavenging peroxy radicals, O_2^-, OH radicals, or reducing oxygen singlet besides obstructing lipid peroxidation (Izunya et al. 2010). A temperature drop during leaves maturation has been verified to boost the effectiveness of flavonoids and phenols (Singh et al. 2008, 2009).

3.4.3 Antioxidant's Mode of Action

Antioxidants possessing low molecular masses (LMWAs) are those small chemicals, which pervade the cells, clump together in abundant amounts in specific compartments connected to oxidation-mediated damage, and subsequently rejuvenate the cell. Cellular LMWAs exist in human tissues. Glutathione, carnosine, and reduced form of NAD (nicotinamide adenine dinucleotide) are formed by cells. Generally, uric acid and bilirubin are produced as metabolic waste products (Ames et al. 1981). Dietary antioxidants include polyphenols, ascorbic acid, and tocopherols. These LMWAs paid specific attention to vitamin C, which is notorious for its reducing assets and is widely employed as an agent of anti-oxidation in food stuffs and beverages. These are important as far as medicines and biological metabolism are concerned.

Ascorbic acid performs the task of a therapeutic antioxidant that aids in immune system activation, wound healing, osteogenesis, detoxification, iron absorption, collagen production, blood vessel clotting prevention, and other innumerable metabolic processes. On exposure to light/heat, vitamin C oxidises quickly, and its breakdown is augmented in the existence of cationic heavy metal. Ascribed to its fluctuating concentration, vitamin C is viewed as a key indicator of food quality that eventually is answerable for the antioxidant characteristics of food (Wawrzyniak et al. 2005).

Low-density lipoproteins are oxidised by a huge number of FRs existing in the host's body, making them increasingly lethal. These might also hasten the ageing process and have been allied to significant illnesses, viz. rheumatoid arthritis, cerebral stroke, cancer, diabetes, and Parkinson and Alzheimer diseases. ROS refer to the molecules with strong oxidising tendency that may be non-radical (O_3, H_2O_2) and/or radical (superoxide and hydroxyl radical) in nature (Campanella et al. 2006). The oxygenated FRs are perceived as vital radical species in biology. Oxidation could be elicited by an extensive range of physico-chemical events, which continues in an appropriate substrate's presence till an obstructive defence mechanism is activated. The targets include oxygen, phospholipids, DNA, cholesterol, and fatty acids mainly polyunsaturated.

3.4.4 Microbial Sources of Antioxidants

Microbial diversity generates a large pool of unique compounds, entailing it a useful cradle for biotechnology innovation. There exist around 23,000 known microbially derived secondary metabolites. Out of 23,000, bacteria including actinomycetes are recognised to produce around 42% of these. Fungi account for 86% of the total, while eubacteria make up the remaining 16%. Microbially derived metabolites have already been substantiated as potent agents for curing bacterial and fungal infections/ diseases (penicillins, amphotericin, streptomycin, erythromycins, vancomycin, and tetracyclines), cancer (daunorubicin, bleomycin, mitomycin, doxorubicins), high cholesterol (lovastatin and mevastatin), and transplant rejection (rapamycin and cyclosporine). Amongst these metabolites, some have antioxidant properties as well (Chandra et al. 2019).

3.4.4.1 Actinomycetes

These are aerobic, filamentous, gram-positive, spore-forming bacteria that produce a large gamut of metabolites accountable for antioxidant, antimicrobial (antifungal and antibacterial), and insecticidal-like biological properties. Mycothiol (MSH), a primary "sugar" thiol found in actinomycetes' cell walls, is a glutathione (GSH) homologue. Actinomycetes lack the enzymes needed to synthesise GSH; hence they rely on low molecular mass thiols like MSH, ergothioneine (ESH), and bacillithiol (BSH). MSH assists in sustaining homeostasis of intracellular redox potential, allowing for the apt working of biological events like DNA synthesis, enzyme activation, and regulation of cell cycle. MSH is an antioxidant, which serves as either an e^- acceptor or donor. It aids in detoxifying xenobiotics and FRs while alkylating agents perform the task of a cofactor. MSH, unlike GSH, is composed of two sugars (inositol and *N*-glucosamine) and amino acids (cysteine, glutamic acid, and glycine).

Streptomyces sp. have been recognised to produce an array of antioxidant isoflavonoids. For instance, metabolites from *S. prunicolor* particularly benthocyanins (A, B, C) and benthophoenin displayed great antioxidant action on liver microsomes of rat in a test method. While compared to synthesised compounds, carbazomycin B and carazostatin (derived from actinomycetes) were revealed to display higher antioxidant potential (Kawahara et al. 2012). The reactive NH and OH group's presence in these molecules could explain their high activity. Carquinostatin A that has both NH and OH groups, and helps to impart protection to the brain, is synthesised by *S. exfoliates.* In microsomes of rat liver, it also displayed antioxidant potential equivalent to vitamin E plus 4′,7,8-trihydroxyisoflavone. These chemicals exhibit anticancer and antioxidant potential. *Streptomyces* LK-3 (JF710608) extract that contains sesamol, gallocatechin gallate, daidzein-8-C-glucoside (puerarin), delphinidin as major component, and cyanidin-3-O-rutinoside has also been acknowledged to exhibit antioxidant potential (Karthik et al. 2013).

Antioxidant potency has been perceived in nitrogen-containing metabolites with NH groups like phenazine hetero and carbazole cycles. Aside from a scarce well-known metabolites like 5-hydroxymaltol obtained from *Streptomyces melanogenes*, and 7-dimethyl naphtherpin plus naphtherin like napthoquinonic derivatives from *S. violaceus* and *S. aeriouvifer*, respectively, have also been discovered. These metabolites prevented lipid peroxidation in the microsomes of rat liver with the value of IC_{50}, which were nearly equal to vitamin E. Stealthins A and B, obtained from *S. viridochromogenes*, had 30 times the potency of vitamin E. Stealthins were more potent than vitamin E as they possess NH, OH, and C=O reactive groups. Brain protecting drugs, flunarizine and BHT, were verified to be substantially less effective than carazostatin generated by *S. chromofuscus.* Neocarzinostatin A to C, obtained from S*treptomyces* spp., were also reported for their potency as an antioxidants (Prashith et al. 2010).

In a nutshell, carbazole chemicals recovered from *Streptomyces* spp. have been acknowledged as a crucial type of antioxidants, which might be beneficial because of being a member of the novel class of therapeutic drugs. *Streptomycetes* spp. produce a distinctive group of antioxidants known as phenazine derivatives, which have all been apparent to have antioxidant activity. For instance, prenylated analogues and benthophoenin were recovered from *S. prunicolor* and *S. exfoliates*, respectively. *S. cyaneus* provides antiostatins from A1 to A4, and B1 to B4, while *S. violaceus* provides carbazoquinocins from A to F, which are carbazole-encompassing *o*-quinone sequences. Benthocyanins like A, B, and C, which are synthesised by *S. prunicolor* have documented to be better in inhibiting lipid peroxidation reaction than vitamin E/tocopherol (Shin-ya et al. 1993). Compounds owning one or/more functional groups inclusive of the OH group may undergo reaction with FRs. Likewise, phenazoviridin, a glycosylated derivative of phenazine, obtained from *Streptomyces* spp. turns as a robust lipid peroxidation inhibitor.

Furthermore, nitrogen-containing antioxidants like benzastatins (A–D) and thiazostatins (A and B) were reported to be synthesised by *S. nitrosporeus* culture broth, which possessed a tetrahydroquinolone ring or an unusual *p*-aminobenzamide unit (Shindo and Misawa 2014). The OH and NH groups are present in benzastatin molecules, while the methoxy group exists in its derivatives. Consequently, it has lower anti-lipid peroxidation potency in the liver microsomes of rat than vitamin E. In N 18-RE-105 (neuroblastoma X-retina hybrid) cell line, however, benzastatins C and D actively help in avoiding glutamate toxicity, signifying that they might be expedient for brain ischaemia injury. The Rhadon-iron method validated that a *Streptomyces*-derived antioxidant molecule, 1,3-Dicarbonyl, has virtually the same action as vitamin C (Rani et al. 2021).

A scanty literature relevant to the antioxidant potency exhibited by marine actinobacteria is available. For instance, *Streptomyces* sp. VITTK3 (marine isolate) displayed a high (96%) DPPH activity with 5 mg/mL concentration. However, 5-2, 4-dimethylbenzyl pyrrolidin-2-one, obtained from another marine isolate, i.e. *Streptomyces* sp., displayed 59.32% DPPH scavenging capacity along with cytotoxic action on oncogenic cells. Correspondingly, there were fewer chromosomal abnormalities in contrast to the control group (Saurav and Kannabiran 2012). Furthermore, JBIR-94 and JBIR-125 like phenolics, recovered from *Streptomyces* spp., have shown 11.4 and 35.1 M of DPPH activity, respectively, having IC_{50} values (Kawahara et al. 2012). Further, the existence of reactive NH and OH groups makes these phenolic compounds effective FR scavengers.

3.4.4.2 Bacteria

Eubacteria are basically prokaryotic microorganisms that could produce an extensive range of extracellular compounds. Bacteria are inextricably linked to all forms of life, including humans, animals, and plants. For instance, probiotic foods, which comprise millions of advantageous bacterial strains, are an indispensable vital component of our diet. Exopolysaccharides (EPS) are powerful antioxidants produced by probiotic bacterial strains that colonise the gastrointestinal system (Rahbar-Saadat et al. 2019). Probiotic/Lactic acid bacterial strains like *Lactobacillus plantarum* and *L. paracasei* subsp. *paracasei* produce enough EPS to demonstrate considerable antioxidant potency (reducing power, DPPH activity, chelation of ferrous ions, and deterrence of linoleic acid peroxidation) with in vitro immunomodulation. *Lactobacillus casei, L. acidophilus*, and *Lactococcus lactis* are three distinct probiotic bacteria thriving in fermented milk and reveal in vivo and in vitro antioxidant potential besides cholesterol assimilation ability. These cultural strains have also revealed potential for DPPH, malondialdehyde, and H_2O_2 scavenging apart from preventing linoleic acid's oxidation.

Moreover, *Lactobacillus casei* demonstrated highest Trolox equivalents (48.7 mM) closely accompanied by *L. acidophilus* (46.3 mM) and *Lactococcus lactis* (23.4 mM) as reported by Jain et al. (2009). Lin and Chang (2000) reported that intact as well as intracellular extracts from intestinal LAB, like *L. acidophilus* and *Bifidobacterium longum*, prevent peroxidation of linoleic acid by its great antioxidant potency. The preliminary screening of 8 isolated *Lactobacillus* strains from Nigerian indigenous fermented foods by Osuntoki for antioxidant potential revealed activity ranging from 2.8 to 31.5% and in a 24-h of fermentation period, selected microbial isolates viz. *L. delbrueckii, L. casei, L. fermentum, L. plantarum*, and *L. brevis* showed considerable DPPH scavenging activity. Wang et al. (2019) reported that EPS was produced as a cell bound entity by *L. fermentum* but as an exogenous product by *L. plantarum* (Min et al. 2019). These exopolysaccharides (EPSs) can be heterogeneous or homogenous, comprised of various sugars like galactose, rhamnose, mannose, arabinose, glucose, xylose, and fructose. Various functional groups, primarily reactive aldehyde, hydroxyl, and ketone groups, were noticed in the structure and chemical configuration of these EPSs. These functional groups possess the capability to react effectively with FRs. Altogether these tested strains displayed radical scavenging capability, with 3–53% inhibition rates. The inhibition rate of *Leuconostoc* and *L. acidophilus* strains varied between 42 and 53% across various strains. Low inhibitory rates of 3–10% were recorded for *Lactobacillus lactis,* and *L. casei* strains plus one kefir strain. Inhibition rates for *Lactobacillus* bacteria ranged from 12 to 42%, while *Lactococcus* strains showed inhibition rates varied from 3 to 22%. Several of these strains were reported to improve significantly the deterrence of lipid peroxidation during fermentation. According to (ABTS) 2,2′-azino-bis (3-ethylbenzothiazoline-6-sulfonic acid), rates of inhibition in fermented milk were over 90%, with whey fractions ranging from 20 to 24%. Furthermore, EPS is frequently synthesised during the practice of fermentation inclusive of other metabolites that could explain why organic acids have a stronger antioxidant effect. For instance, in rats, daily inclusion of *L. casei* (28×10^{11} CFU/rat) for up to 8 days declined the OS and liver lesions induced consequently upon introduction of carbon tetrachloride single intraperitoneal (Liu et al. 2012). Intracellular metabolites derived from *L. casei* were reported to possess the capability of conquering aflatoxin B1 produced in rats due to OS (Aguilar-Toalá et al. 2019).

Similarly, *Pseudomonas* spp. are reported to yield a resorcinol-type alkyloid compound, resorstatin, with two reactive OH groups that possesses anti-lipoperoxidative properties analogous to BHT. The xanthin synthesised by this bacterium also repressed linoleic acid oxidation, which acted in concurrence with tocopherol (Kato et al. 2012). The composition of epiphytic bacteria allied with *Bifurcaria bifurcate* (marine brown alga) was revealed by Horta et al. (2014) who described that *Vibrio* sp. makes up over half (48.72%) of this bacterial population. Other important bacteria in this community included *Alteromonas* sp. (12.82%) and *Shewanella* sp. (12.26%) besides *Serratia* sp., *Staphylococcus* sp., *Citricoccus* sp., *Ruegeria* sp., and *Cellulophaga* sp. which constitute same content of 2.5%. The capability of these bacterial strains to be a virtuous source of naturally produced antioxidants was appraised by computing their O_2 radical absorbance capacity (ORAC), DPPH activity, and total phenols in solvent extracts made using methanol and dichloromethane (Horta et al. 2014). Likewise, endophytic bacterial strains, viz. *Bacillus anthracis, Pseudomonas hibiscicola, Enterobacter ludwigii*, and *Macrococcus caseolyticus* isolated from *Aloe vera*, have been recognised to synthesize bioactive molecules possessing pronounced DPPH (75–88%) activity. Plants can develop even when they are stressed owing to the existence of these antioxidants (Akinsanya et al. 2015).

3.4.4.3 Blue Green Algae

Photosynthetic eubacteria include photosynthetic bacteria that are also acknowledged as blue green algae (BGA) or cyanobacteria. This cluster of microorganisms is rich in carotenoids, viz. carotene, lutein, and lycopene. Carotenoids have natural antioxidant and anti-inflammatory potential as they are good at quenching ROS. BGA also contain phycocyanin-like pigments in great conc., around 15% on dry wt. basis. C-phycocyanin is a potent hepatoprotector and scavenger of FRs. Phycocyanin, a water soluble molecule with N-H reactive groups, has also revealed reduction in inflammation rate in mice ears and protects rats from colitis caused by acetic acid.

Another novel carotenoid pigment, astaxanthin, has been documented to perform an array of metabolic functions like protection from oxidation and UV radiation from the sun, improved eyesight, immunological response, and pigmentation formation. Various in vitro investigations have detected and noted its potential to quench oxygen singlet and FRs which was roughly ten times that of carotenoids and multiple times that of tocopherol (Dose et al. 2016). Another microalgal isolate, *Haematococcus pluvialis*, has a high astaxanthin content and might be engaged in diverse applications (Naguib 2000). Catechin and quinic acid-like phenolic compounds with strong antioxidant potential have been documented to be synthesised by *Arthrospira, Nostoc,* and *Anabaena* (Blagojević et al. 2018). Phycobili protein, a cyanobacteria's metabolite, is a highly valued natural product, whose usage can be made in a varied number of biotechnological products. It contains OH, NH, COOH, and C=O like reactive functional groups, which help in lessening ROS formation. Fabrication of these cyanobacterial phycobiliproteins necessitates an optimal medium, and several bioprocesses influence bacteria growth besides the amassing of phycobili proteins (Pagels et al. 2019).

3.4.4.4 Lichens

Algae and fungi form a symbiotic connotation known as lichen. As a product of their relationship, these species produce an array of extracellular secondary metabolites. The majority of these were found only in lichens that may be exploited as natural antioxidants (Kosanić and Ranković 2015). Nevertheless, there is paucity of research regarding examining the antioxidative potency of lichens. In vitro antioxidant potency displayed by the aqueous extracts of *Cetraria islandica* was originally demonstrated by Gülçin et al. (2002) who observed that the antioxidant potency of this organism was far greater than tocopherol. Other lichens with high methanolic content, such as *Umbilicaria nylanderiana, Ramalina pollinaria, R. polymorpha, Platismatia glauca,* and *Parmelia saxatilis*, prevented linoleic acid's oxidation and scavenged free radicals (Gulluce et al. 2006).

Three different approaches, viz. reducing power plus DPPH and superoxide anion radical (O_2^-) scavenging activities, were employed by Ranković (2011) for evaluating the antioxidant potency exhibited by diverse lichens (*Umbilicaria cylindrical, Lecanora atra, Cetraria islandica, Pseudoevernia furfuraceae,* and *Parmelia pertusa*) extracts prepared using different solvents like aqueous, methanol and acetone. The extracts displayed DPPH activity from 32 to 95%, reducing power values from 0.016 to 0.109, while O_2^- scavenging activity varied between 7 and 84%. Total phenols (12–76.42 g of pyrocatechol equiv.) and total flavonoids (1.37–54.77 g of rutin equiv.) were found in great quantities in these tested extracts, signifying phenols plus flavonoids as fundamental antioxidant components (Ranković 2011). This diversity in lichens ensures novel bioactive molecule's synthesis with significant antioxidant potential, making them a natural antioxidant's source. Purification of usnic acid, praesorediosic acid, collatolic acid, chloroatranorin, and protocetraric acid with methanol and acetone extracts from lichens for example *Parmotrema rampoddense, P. praesorediosum, P. tinctorum,* and *P. reticulatum* revealed their phenolic nature. The DPPH protocol demonstrated that these molecules exhibit a considerable antioxidant potential (Rajan et al. 2011). Diverse range of compounds viz. protolichesterinic acid, atranorin, usnic acid, sekikaic acid, 2-hydroxy-4-methoxy-6-propyl benzoic acid, benzoic acid, homosekikaic acid, and 2,4-dihydroxy-6-propyl were noticed in hexane extracts (nonpolar solvent) of lichens. Among these molecules, sekikaic acid exhibits the greatest DPPH activity, accompanied by homosekikaic acid. Other persuasive antioxidants like montagnetol, orcinol, methyl haematomate, orsellinic acid, *p*-depsides, and methyl-orcinolcarboxylate (e.g. divericatic acid, atranorin, erythrin, and lecanoric acid) have also been recovered from *H. obscurata, R. montagnei, Cladonia* sp., and *P. grayana* (Sisodia et al. 2013). *Pleurosticta acetabulum* was observed to scavenge DPPH radicals (IC50 151 g/mL). Evernic acid, salazinic, atranorin, protocetraric, and norstictic were noticed as the active components of this particular lichen. These compounds possess antibacterial plus anticancer properties besides being antioxidants (Tomović et al. 2017). These molecules usually have a wide array of reactive functional group whose location affects their antioxidant capability.

3.4.4.5 Fungi

Since the dawn of civilisation, fungi have aided the well-being of humanity. There exist nearby 1.5 million fungal species in the fungi kingdom, which possesses a diverse variety. In their interactions with people, fungi are both useful and toxic, albeit most of them are benign. Flavonoids, tetralones, alkaloids, steroids, benzoquinones, organic acids, terpenoids, phenols, xanthones, and other primary and secondary metabolites may be synthesised by an enormous number of fungal groups. These metabolites perform diverse biological activities like anti-oxidation and their molecular structure influences their function substantially.

Filamentous mushrooms of various phyla are infrequently included in human diets, yet they are documented to contain antioxidants. Citric acid is produced primarily by *Aspergillus niger* strains-mediated fermentation, which are a worthy source of ascorbic acid. Food additives, viz. emulsifiers, stabilisers, and flavours, as well as lipases, dietary fibres, poly- and oligosaccharides, triterpenoids, proteins, peptides, phenols, minerals, alcohols, and vitamins, are all active and valuable medical goods. Numerous phenolic acid derivatives, synthesised by *Chaetomium* spp., *Cladosporium* spp., *Phoma* spp., and *Torula* sp., along with secondary metabolites viz. terpenoids, rutin, and benzoic acid. Rutin, phenolic acid's derivatives, and chlorogenic acid are documented to exhibit an extensive range of antibacterial, immunomodulatory, antimutagenic, antioxidant, and antiviral-like biological properties (Huang et al. 2007).

Several filamentous species of fungal genera viz. *Torula*, *Penicillium*, *Aspergillus*, *Eurotium*, *Colletotrichum*, *Mortierella*, *Phoma*, and *Cladosporium* have been reported to produce antioxidant metabolites (Huang et al. 2007). Fungal metabolites were validated to possess a stronger free radical scavenging potential when equated to synthetic antioxidants and phytochemicals. The phenolic glycoside's potency to scavenge FRs is mostly on account of their phenolic and hydroxyl groups. *Pestalotiopsis microspora* (an endophytic fungus) synthesised both antifungal and antioxidant chemicals like 1,3-dihydroisofuran, isopestacin, and pestacin, whose activities surpassed vitamin E by greater than tenfold. Isopestacin is assumed to exhibit antioxidant potential owing to its structural similarities with flavonoids. ESRS (electron spin resonance spectroscopy) affirmed the compound's capability to scavenge O_2^- and OH like FRs from the solution.

In India, particularly from Punjab region, 120 different fungal species, recovered from soil, have been examined by Chandra and Arora (2017) for their antioxidant behaviour employing dot blot assay technique. Results shown that out of 120, only 51 isolates displayed antioxidant potential, hence were assayed further quantitatively (Chandra and Arora 2017). Various soil fungal strains viz. *Penicillium expansum, Aspergillus fumigatus, A. terreus, Penicillium granulatum, A. wentii,* and *P. citrinum* revealed potential for antioxidant optimization (Chandra and Arora 2012). Arora et al. (2011) discovered that fungi responsible for wood degradation, such as *Phlebia brevispora, Phanerochaete chrysosporium, Phlebia radiata, Daedalea flavida, Phlebia floridensis, Phlebia fascicularia,* and *Ceriporiopsis subvermispora*, exhibit antioxidant activity besides depolymerizing lignin found in agricultural biomass into numerous phenol-derivative compounds that might be exploited as animal fodder with enhanced antioxidant potential.

Kojic acid is a secondary metabolic product of *Penicillium* and *Aspergillus.* This chemical obstructs tyrosinase activity and might be utilised as a food additive besides melasma therapeutic, antioxidant, antitumor, and radioprotective agent. Derivatives of kojic acid are recognised to exert cytotoxic and anti-proliferative effects in vitro (Kono et al. 2001). A chlorine-encompassing pigment, sclerotiorin, synthesised by *Penicillium frequentans* and *P. sclerotiorum*, inhibited soybean lipoxygenase-1 (LOX-1) in a reversible and non-competitive manner. This inhibitor also displays antioxidant properties as it has the ability to scavenge FRs besides inhibiting lipid peroxidation (non-enzymatic). Literature sources reveal that sclerotiorin inhibits LOX via reaction with ES (enzyme-substrate) complex and scavenging the FR's intermediates synthesised during enzyme activity, and its activity was analogous to both natural and synthetic lipoxygenase inhibitors. Sclerotiorin is basically an azaphilone, a kind of compound with a characteristic isochromane ring structure (Tabata et al. 2012). Further, these fungi's antioxidant properties promote their symbiotic/mutualistic relationship with plant species that may help the plants battle OS triggered by environmental influences.

3.4.4.6 Yeast

During fermentation, yeast produces antioxidants, which boost the approachability of therapeutic molecules and antimicrobial compounds, which limit the growth of moulds and pathogenic bacteria. Yeast microflora thrive in raw materials, food items (cheese, yoghurt, and sausage), fruits, and meat. Rai et al. (2021) confirmed the antioxidant potency of yeasts, viz. *Pichia kudriavzevii, Kluyveromyces marxianus, Candida utilis* and *Yarrowia lipolytica*. *Saccharomyces cerevisiae* strains, which have been genetically engineered, were observed to be more efficient at producing the antioxidant chemicals, resveratrol and astaxanthin. Glucan and D-mannan, present in yeast's cell wall, contain a vast amount of OH groups accountable for high metal chelation and reducing capability (Al-Manhel and Niamah 2017).

Fakhruddin et al. (2017) demonstrated the production of astaxanthin from *Phaffia rhodozyma* (red yeast). The antioxidant of this yeast is ascribed to the existence of β-D-glucans (β-DG), polysaccharides and α-D-mannans (α-DM) in its cell wall. β-DG and α-DM demonstrated a protective impact of an aqueous-soluble carboxymethylated yeast (CMG) both in phosphatidylcholine liposomes and lipid peroxidation. CMG inhibits peroxidation caused by ultraviolet A (UVA) that led to singlet oxygen's formation. The European Food Safety Authority (EFSA) has approved the usage of β-glucans produced by *Saccharomyces* spp. The usage of β-glucans producing *Saccharomyces* spp. has been permitted by EFSA. The β-glucans of *Saccharomyces* spp. were also approved as food additives (EFSA 2011).

Peroxidases are widely expressed enzymes that convert H_2O_2 and alkyl hydroperoxides to H_2O or alcohol. Based on conserved 1 or 2 cysteine residues, the Prx superfamily is categorised into two subtypes: 1-Cys Prx and 2-Cys Prx. Prx was the first to uncover a protein in *S. cerevisiae* that might inhibit glutamine synthetase from reactivating through the thiol/Fe(III)/oxygen oxidation pathway system. The *S. cerevisiae's* genome encompasses an "ORF YBL064C" sequence, which is reported to code for a protein similar to 1-Cys Prx protein of humans. Yeasts have two Trx1 and Trx2 like cytosolic thioredoxins. Both are replenished through cytosolic regeneration. Different oxidoreductases (thiol-dependent) derived from *S. cerevisiae*, glutathione peroxidases (Gpx1 to Gpx3) and peroxiredoxins (Tsa2, Ahp1, Tsa1, Prx1, Dot5), detoxify peroxides by connecting a disulphide link between resolvAhp and peroxidase. Tsa1 is a multitasking protein that provides protection to the cell against OS triggered by aggregation and misfolding of nascent-proteins through behaving like a redox switch, protein chaperone, and a particular antioxidant (Garrigós et al. 2020).

3.5 Conclusion

Microorganisms can be envisaged as worthy and economical sources of naturally produced antioxidant molecules as these are easy to grow and yield antioxidants in a shorter period of time. Microbially derived antioxidants are non-mutagenic and non-cytotoxic in nature in comparison to synthetic ones. An enormous microflora like microalgae, bacteria, actinomycetes, and yeast has been recognised as potent antioxidants producers. Molecules like pestacin, polysaccharides, astaxanthin, and isopestacin possess good antioxidant potency and are thus employed as functional components in numerous pharmaceuticals, cosmetics, food stuffs, and nutraceutical products. Glutathione and ergothioneine are abundant in glutathione and ergothioneine-rich cultivated and wild mushrooms, which aid in avoiding mitochondrial damage instigated by O_2^- FRs. Bioprospecting of chemicals found in *Saccharomyces* sp., such as torularhodin, thioredoxins, and peroxiredoxins, opens up a new boulevard for yeast to be exploited as an antioxidant source plus in dietary add-ons and other industrial applications. In a nutshell, an in-depth knowledge of antioxidants' mode of action, structures, optimization, and other aspects may assist in the formation of novel medicines and other sectors using the enormous biodiversity of microorganisms.

Acknowledgement The authors are highly thankful to Mr. Eklavya Chauhan for providing assistance in making figures.

References

Aguilar-Toalá JE, Estrada-Montoya MC, Liceaga A et al (2019) An insight on antioxidant properties of the intracellular content of *Lactobacillus casei* CRL-431. LWT 10:1016

Akinsanya M, Goh J, Ping L et al (2015) Diversity, antimicrobial and antioxidant activities of culturable bacterial endophyte communities in Aloe vera. FEMS Microbiol Lett 362:10

Al-Manhel AJ, Niamah AK (2017) Mannan extract from *Saccharomyces cerevisiae* used as prebiotic in bio-yogurt production from buffalo milk. Int Food Res J 24:2259–2264

Amany M, Abdel-Raheam HE (2020) Red and yellow Monascus pigments as potential natural antioxidants for fatty foods. Plant Arch 20:444–449

Ames BN, Cathcart R, Schwiers E et al (1981) Uric acid provides an antioxidant defense in humans against oxidant and radical-caused aging and cancer: a hypothesis. Proc Natl Acad Sci USA 78: 6858–6862

Anwar H, Hussain G, Mustafa I (2018) Antioxidants from natural sources. Antiox Food Appl 3:28

Arias A, Feijoo G, Moreira MT (2022) Exploring the potential of antioxidants from fruits and vegetables and strategies for their recovery. IFSET 4:102974

Arora DS, Sharma R, Chandra P (2011) Biodelignification of wheat straw and its effect on in vitro digestibility and antioxidant properties. Int Biodeter Biodegrad 65:352–358

Barbalho SM, Direito R, Laurindo LF et al (2022) Ginkgo biloba in the aging process: a narrative review. Antioxidants (Basel) 11:525

Beer DG, Kardia SL, Huang CC et al (2002) Gene-expression profiles predict survival of patients with lung adenocarcinoma. Nat Med 8:816–824

Bensid A, El Abed N, Houicher A et al (2022) Antioxidant and antimicrobial preservatives: properties, mechanism of action and applications in food—a review. Crit Rev Food Sci Nutr 62:2985–3001

Blagojević K, Babić O, Raseta M et al (2018) Antioxidant activity and phenolic profile in filamentous cyanobacteria: the impact of nitrogen. J Appl Phycol 30:10

Boveris A, Repetto M, Bustamante J, Boveris AD, Valdez LB (2006) The concept of oxidative stress in pathology. Free Radic Pathophysiol 10:1–17

Campanella L, Favero G, Persi L et al (2006) New biosensor for superoxide radical used to evidence molecules of biomedical and pharmaceutical interest having radical scavenging properties. J Pharm Biomed Anal 23:10

Cartea M, Francisco M, Soengas P et al (2011) Phenolic compounds in brassica vegetables. Molecules (Basel, Switzerland) 16:251–280

Chambial S, Dwivedi S, Shukla KK, John PJ, Sharma P (2013) Vitamin C in disease prevention and cure: an overview. Indian J Clin Biochem 28(4):314–328

Chandra P, Arora DS (2012) Optimization of antioxidant potential of *Penicillium granulatum* Bainier by statistical approaches. ISRN Microbiol 2012:452024

Chandra P, Arora DS (2017) Antioxidant potential of *Penicillium expansum* and purification of its functional compound. Asian J Biotechnol 9:24–34

Chandra P, Sharma R, Arora DS (2019) Antioxidant compounds from microbial sources: a review. Int Food Res J 129:108849

Chandra P, Sharma RK, Arora DS (2020) Antioxidant compounds from microbial sources: a review. Food Res Int 129:108849

Dembinska-Kiec A, Mykkänen O, Kiec-Wilk B et al (2008) Antioxidant phytochemicals against type 2 diabetes. Br J Nutr 99:109–117

dePascual-Teresa S, Moreno DA, García-Viguera C (2010) Flavanols and anthocyanins in cardiovascular health: a review of current evidence. Int J Mol Sci 11:1679–1703

Dose J, Matsugo S, Yokokawa H et al (2016) Free radical scavenging and cellular antioxidant properties of astaxanthin. Int J Mol Sci 17:103

European Food Safety Authority (2011) Evaluation of the food, the food classification system applied to the development of the EFSA Comprehensive European Food Consumption Database. EFSA J 9:1970

Fakhruddin S, Alanazi W, Jackson KE (2017) Diabetes-induced reactive oxygen species: mechanism of their generation and role in renal injury. J Diabetes Res 2017:8379327

Freeman BA, Crapo JD (1982) Biology of disease: free radicals and tissue injury. Lab Investig 47: 412–426

García-García R, Searle SS (2016) Preservatives: food use. In: Caballero B, Finglas PM, Toldrá F (eds) Encyclopedia of food and health. Academic, New York, pp 505–509. https://doi.org/10.1016/B978-0-12-384947-2.00568-7

Garrigós MC, Mellinas AC, Jiménez A (2020) Optimization of microwave-assisted extraction of cocoa bean shell waste and evaluation of its antioxidant, physicochemical and functional properties. LWT 127:109361

Gassen M, Youdim MB (1999) Free radical scavengers: chemical concept and chemical relevance. J Neural Transm Suppl 56:193–210

Glenville J, Byford V, West S et al (2006) Hepatic activation and inactivation of clinically-relevant vitamin D analogs and prodrugs. Anticancer Res 26:2589–2595

Gulcin I (2020) Antioxidants and antioxidant methods: an updated overview. Arch Toxicol 94:651–715

Gülçin I, Oktay M, Küfrevioğlu OI et al (2002) Determination of antioxidant activity of lichen *Cetraria islandica* (L) Ach. J Ethnopharmacol 79:325–329

Gulluce M, Aslan A, Sokmen M et al (2006) Screening the antioxidant and antimicrobial properties of the lichens *Parmelia saxatilis, Platismatia glauca, Ramalina pollinaria, Ramalina polymorpha* and *Umbilicaria nylanderiana*. Phytomedicine 13:515–521

Han H, Yılmaz H, Gülçin I (2018) Antioxidant activity of flaxseed (Linum usitatissimum L.) shell and analysis of its polyphenol contents by LC-MS/MS. Rec Nat Prod 12:397–402

Horta A, Pinteus S, Alves C et al (2014) Antioxidant and antimicrobial potential of the *Bifurcaria bifurcata* epiphytic bacteria. Mar Drugs 12:1676–1689

Huang JC, Li DJ, Diao JC et al (2007) A novel fluorescent method for determination of peroxynitrite using folic acid as a probe. Talanta 72:1283–1287

Igarashi K, Kashiwagi K (2000) Polyamines: mysterious modulators of cellular functions. Biochem Biophys Res Commun 271:559–564

Izunya AM, Nwaopara AO, Aigbiremolen A, Odile MAC, Oaikhena GA, Bankole JK, Ogarah PA (2010) Morphological and biochemical effects of crude aqueous extract of *Mangifera indica* L. (mango) stem bark on the liver in Wistar rats. Res J Appl Sci Eng Technol 2:460–465

Jain D, Daima H, Kachhwala S et al (2009) Synthesis of plant-mediated silver nanoparticles using papaya fruit extract and evaluation of their anti-microbial activities. Dig J Nanomater Biostruct 4:557–563

Jîtca G, Osz BE, Tero-Vescan A et al (2022) Positive aspects of oxidative stress at different levels of the human body: a review. Antioxidants 11:572

Karthik L, Kumar G, Rao KV (2013) Antioxidant activity of newly discovered lineage of marine actinobacteria. Asian Pac J Trop Med 6:325–332

Kato H, Takemoto D, Kawakita K (2012) Proteomic analysis of S-nitrosylated proteins in potato plant. Physiol Plant 148:371–386

Kawahara Y, Oono Y, Kanamori H et al (2012) Simultaneous RNA-Seq analysis of a mixed transcriptome of rice and blast fungus interaction. PLoS One 7:e49423

Kono H, Rusyn I, Uesugi T et al (2001) Diphenyleneiodonium sulfate, an NADPH oxidase inhibitor, prevents early alcohol-induced liver injury in the rat. Am J Physiol Gastrointest Liver Physiol 280:G1005–G1012

Kosanić M, Ranković B (2015) Studies on antioxidant properties of lichen secondary metabolites. In: Ranković B (ed) Lichen secondary metabolites bioactive properties and pharmaceutical potential. Springer International Publishing, New York, pp 105–125

Kumar M, Kumar S, Kaur S (2011) Investigations on DNA protective and antioxidant potential of chloroform and ethyl acetate fractions of *Koelreuteria paniculata Laxm.* Afr J Pharm Pharmacol 5:421–427

Kumar S, Chaitanya RK, Preedy VR (2018) Assessment of antioxidant potential of dietary components. In: HIV/AIDS. Academic Press, New York, pp 239–253

Li S, Li H, Xu X (2021) Nanocarrier-mediated antioxidant delivery for liver diseases. Theranostics 10:1262–1280

Lin MY, Chang FJ (2000) Antioxidative effect of intestinal bacteria *Bifidobacterium longum* ATCC 15708 and *lactobacillus acidophilus* ATCC 4356. Dig Dis Sci 45:1617–1622

Liu J, Wisniewski M, Droby S et al (2012) Increase in antioxidant gene transcripts, stress tolerance and bio-control efficacy of *Candida oleophila* following sublethal oxidative stress exposure. FEMS Microbiol Ecol 80:578–590

Manach C, Scalbert A, Morand C (2004) Polyphenols: food sources and bioavailability. Am J Clin Nutr 79:727–747

Min WH, Fang XB, Wu T et al (2019) Characterization and antioxidant activity of an acidic exopolysaccharide from *Lactobacillus plantarum* JLAU103. J Biosci Bioeng 127:758–766

Moussa Z, Judeh ZM, Ahmed SA (2019) Nonenzymatic exogenous and endogenous antioxidants. In: Free radical medicine and biology. IntechOpen, London, pp 2631–8261

Naguib YM (2000) Antioxidant activities of astaxanthin and related carotenoids. J Agric Food Chem 48:1150–1154

Ondrej H, Mezeyova I, Hegedusova A, Vargova A, Timoracka M, Slosar M, Andrejiova A, Jurikova T, Mezey J (2020) Basil seeds as a source of antioxidants affected by fortification with selenium. Folia Horticult 32:11–20

Pagels F, Guedes AC, Amaro HM et al (2019) Phycobiliproteins from cyanobacteria: chemistry and biotechnological applications: chemistry and biotechnological applications. Biotechnol Adv 37: 422–443

Parker DS, Blauwkamp T, Cadigan K et al (2007) Wnt/β-catenin-mediated transcriptional regulation. Adv Dev Biol 17:1–60

Patel J (2008) A review of potential health benefits of flavonoids. Lethbridge Undergraduate Res J 3(2):1–5

Pimpa B, Kanjanasopa D, Boonlam S (2009) Effect of addition of antioxidants on the oxidative stability of refined bleached and deodorized palm olein. J Kasetstart Nat Sci 43:370–377

Rahbar-Saadat Y, Yari AK, Pourghassem GB (2019) A comprehensive review of anticancer, immunomodulatory and health beneficial effects of the lactic acid bacteria exopolysaccharides. Carbohydr Polym 217:79–89

Rai R, Singh S, Rai KK (2021) Regulation of antioxidant defense and glyoxalase systems in cyanobacteria. Plant Physiol Biochem 168:353–372

Rajan S, Gokila M, Jency P et al (2011) Antioxidant and phytochemical properties of *Aegle marmelos* fruit pulp. Int J Curr Pharm Res 3:65–70

Rani A, Saini KC, Bast F et al (2021) Microorganisms: a potential source of bioactive molecules for antioxidant applications. Molecules 26:1142

Ranković B (2011) Antioxidant properties of some lichen species. J Food Sci Technol Mysore 48: 584–590

Sasse A, Colindres P, Brewer MS (2009) Effects of natural and synthetic antioxidants on the oxidative stability of cooked, frozen pork patties. J Food Sci 74:30–35

Saurav K, Kannabiran K (2012) Cytotoxicity and antioxidant activity of 5-(2,4-dimethylbenzyl) pyrrolidin-2-one extracted from marine *Streptomyces* VITSVK5 spp. Saudi J Biol Sci 19:81–86

Sdona E, Ekström S, Andersson N et al (2022) Fruit, vegetable and dietary antioxidant intake in school age, respiratory health up to young adulthood. Clin Exp Allergy 2:104–114

Shati AA, Zaki MS, Alqahtani YA et al (2021) Potential protective effect of vitamin C on Qunalphos-induced cardiac toxicity: histological and tissue biomarker assay. Biomedicine 10: 39

Shindo K, Misawa N (2014) New and rare carotenoids isolated from marine bacteria and their antioxidant activities. Mar drugs 12:1690–1698

Shin-ya K, Tanaka M, Furihata K, Hayakawa Y, Seto H (1993) Structure of carquinostatin A, a new neuronal cell protecting substance produced by *Streptomyces exfoliatus*. Tetrahedron Lett 34: 4943–4944

Shukla S, Mehta A, Bajpai VK, Shukla S (2009) *In vitro* antioxidant activity and total phenolic content of ethanolic leaf extract of *Stevia rebaudiana* Bert. Food Chem Toxicol 47:2338–2343

Sies H (1985) Oxidative stress: introductory remarks. In: Sies H (ed) Oxidative stress. Academic, London, pp 1–8

Singh P, Kumar R, Sabapathy SN et al (2008) Functional and edible uses of soy protein products. Compr Rev Food Sci Food Saf 7:14–28

Singh BN, Singh BR, Singh RL et al (2009) Polyphenolics from various extracts/fractions of red onion (*Allium cepa*) peel with potent anti-oxidant and anti-mutagenic activities. Food Chem Toxicol 47:1161–1167

Singh J, Solomon S, Kumar D (2013) Manufacturing Jaggery, a product of sugarcane, as health food. Agrotechnology S11:7. https://doi.org/10.4172/2168-9881.S11-007

Singh RL, Sharma S, Singh P (2014) 16 Antioxidants: their health benefits and plant sources. In: Phytochemicals of nutraceutical importance, vol 28. CABI, Wallingford, pp 248–265

Singh B, Singh JP, Kaur A, Singh N (2017) Phenolic composition and antioxidant potential of grain legume seeds: a review. Food Res Int 101:1–6

Sisodia R, Geol M, Verma S et al (2013) Antibacterial and antioxidant activity of lichen species *Ramalina roesleri*. Nat Prod Res 27:2235–2239

Solomons NW (2001) Vitamin A and carotenoids. In: Present knowledge in nutrition, 8th edn. ISLI Press, Washington, DC

Prashith TR, Shobha KS, Onkarappa R (2010) Studies on antioxidant and anthelmintic activity of two Streptomyces species isolated from Western Ghat soils of Agumbe, Karnataka. J Pharm Res 3:26–29
Tabata S, Ikeda R, Yamamoto M et al (2012) Thymidine phosphorylase enhances reactive oxygen species generation and interleukin-8 expression in human cancer cells. Oncol Rep 28:895–902
Timoshnikov VA, Selyutina OY, Polyakov NE et al (2022) Mechanistic insights of chelator complexes with essential transition metals: antioxidant/pro-oxidant activity and applications in medicine. Int J Mol Sci 23:1247
Tomović V, Jokanović M, Branislav Š et al (2017) Plants as natural antioxidants for meat products. IOP Conf Ser Earth Environ Sci 85:012030
Tsuda T, Horio F, Kitoh J, Osawa T (1999) Protective effects of dietary cyaniding 3-O-β-D-glucoside on liver ischemia-reperfusion injury in rats. Arch Biochem Biophys 368:361–366
Tsuda T, Kato Y, Osawa T (2000) Mechanism for the peroxynitrite scavenging activity by anthocyanins. FEBS Lett 484:207–210
Wang S, Qian K, Zhu Y et al (2019) Reactivation and pilot-scale application of long-term storage denitrification biofilm based on flow cytometry. Water Res 148:368–377
Wawrzyniak A, Marciniak A, Rajewska J (2005) Lycopene content of selected foods available on the polish market and estimation of its intake. Pol J Food Nutr Sci 14:55
Wolf G (2005) The discovery of the antioxidant function of vitamin E: the contribution of Henry A. Mattill. J Nutr 135:363–366
Wrigstedt P, Kylli P, Pitkanen L et al (2010) Synthesis and antioxidant activity of hydroxycinnamic acid xylan esters. J Agric Food Chem 58:6937–6943
Wu B, Cui H, Xi P et al (2013) Investigation of the serum oxidative stress in broilers fed on diets supplemented with nickel chloride. Health 5:454–459
Young AJ, Lowe GL (2018) Carotenoids—antioxidant properties. Antioxidants 7:28

Chapter 4
Microbial and Bio-based Preservatives: Recent Advances in Antimicrobial Compounds

Charles Obinwanne Okoye, Emmanuel Sunday Okeke, Timothy Prince Chidike Ezeorba, Kingsley Ikechukwu Chukwudozie, Chike Obinna Chiejina, and Nidelle Sausten Fomena Temgoua

Abstract The biodeterioration of food products remains a significant public health concern, and ensuring a safe food supply is a substantial challenge for the food industry. The antimicrobial compounds synthesized during fermentation have demonstrated inhibitory effects against many foodborne pathogenic microorganisms. Recently, investigations have been carried out globally to develop safe, natural

C. O. Okoye
Department of Zoology & Environmental Biology, Faculty of Biological Sciences, University of Nigeria, Nsukka, Nigeria

Biofuels Institute, School of Environment & Safety Engineering, Jiangsu University, Zhenjiang, China

E. S. Okeke
Institute of Environmental Health & Ecological Security, School of Environment & Safety Engineering, Jiangsu University, Zhenjiang, China

Natural Science Unit, School of General Studies, University of Nigeria, Nsukka, Nigeria

Department of Biochemistry, Faculty of Biological Sciences, University of Nigeria, Nsukka, Nigeria

T. P. C. Ezeorba
Department of Biochemistry, Faculty of Biological Sciences, University of Nigeria, Nsukka, Nigeria

K. I. Chukwudozie (✉)
Department of Microbiology, Faculty of Biological Sciences, University of Nigeria, Nsukka, Nigeria

Department of Clinical Medicine, School of Medicine, Jiangsu University, Zhenjiang, China
e-mail: ikechukwu.chukwudozie@unn.edu.ng

C. O. Chiejina
Department of Zoology & Environmental Biology, Faculty of Biological Sciences, University of Nigeria, Nsukka, Nigeria

N. S. Fomena Temgoua
School of Food & Biological Engineering, Jiangsu University, Zhenjiang, China

A. K. Nadda, G. Goel (eds.), *Microbes for Natural Food Additives*, Microorganisms for Sustainability 38, https://doi.org/10.1007/978-981-19-5711-6_4

food preservatives to protect food products, and advances have been made to meet users' acceptance as a substitute for chemical preservatives. Antimicrobial compounds could provide an innovative and exciting benchmark with the burgeoning application of natural and biological preservatives in the food industry. Thus, this chapter provides an overview of microbial and bio-based preservatives as a potential transformation of natural preservatives in meeting the needs of the food industry.

Keywords Microbial preservatives · Bio-based preservatives · Antimicrobial compounds · Food products

4.1 Introduction

Preservatives are food additives that protect food from deterioration caused by fungi, bacteria, and other microbes and prolong shelf life by minimizing degradation and other undesirable responses (Fan et al. 2018). Salt and sugar are two of the oldest preservatives, as they preserve foods by lowering their water activity to the point where bacteria cannot grow. Vinegar, on the other hand, preserves by reducing the pH. These can be used on their own or combined with other food preservation procedures, including smoking and pasteurization (Tomaska and Brooke-Taylor 2014). Antimicrobials, which prevent bacteria and fungi from growing on food, and antioxidants, which prevent food ingredients from oxidizing, are two main preservatives. Chemical food preservatives, including benzoic acid, nitrites, butylated hydroxyanisole, and butylated hydroxytoluene, have been extremely beneficial to humanity. Although most chemical preservatives are now deemed safe, there have been concerns about the safety of some of these substances (Aluyor and Oboh 2014).

Consumers have expressed misgivings about the safety and use of food preservatives in recent years (Chen and Nielsen 2016). Researchers have been developing novel natural preservative systems to address the need for natural product formulations and give alternatives to bio-based preservatives. This chapter aims to discuss the current understanding and applications of bio-based antimicrobial preservatives, such as plant, animal, and microorganisms, as well as their mechanisms of action. In addition, recent advances in antimicrobial preservatives, including the production of organic acids from engineered microbes for large-scale industrial applications, were elucidated.

4.2 Bio-based Preservatives and Their Mechanisms of Action

Bio-based food preservatives generally deter spoilage organisms and pathogens from food and food products to achieve food preservation and consumers' protection (Amit et al. 2017). Food preservation ensures the quality (both the sensory and organoleptic properties as well as the safety profile) of the food is maintained for an extended time—both in and out of season (Gokoglu 2019; Torres-León and Aguilar 2022). Up to date, countless antimicrobial agents have been discovered and reported with varying potency levels (Novais et al. 2022). These antimicrobials could be classified differently based on their nature and sources, mode of action, or even their adverse effects or toxicity level. Specifically, based on their nature and sources, antimicrobials are either natural products (biological preservatives), synthetic substances (chemical preservatives), or even physical treatments (Dey and Nagababu 2022). An optimal antimicrobial preservative should at least meet the following criteria: first, maintain or improve the sensory and organoleptic properties as well as the nutritional value of the food; second, foster the microbial protection against foodborne pathogens and anti-nutritional microbial activities; and finally, does not constitute any form of toxicity—whether acute or chronic (Torres-León and Aguilar 2022).

Based on the abovementioned criteria, natural products or biopreservatives have been reported in many studies as the choicest candidates, compared to synthetic chemical preservatives or even physical treatments (Bensid et al. 2020). Bio-based antimicrobial products (BIOBAM) are highly and selectively potent in mitigating the activities of a broad spectrum of food spoilage microbes and pathogens, with negligible toxicities to humans. Moreover, some natural products improve food's organoleptic and nutritional properties. Natural or BIOBAM are classified based on their sources—plant-derived, animal-derived, and microorganism-derived BIOBAM (El-Saber Batiha et al. 2021). Also, the biopreservative and bio-protection activities of BIOBAMs could be through one of the following mechanisms: first, inhibition of the activities of ATPase; second, membrane rupture causing leakage of biomolecules; third disrupting the microbial proton balance or electromotive force; and finally, by enzyme inactivation, inhibition, or denaturation (Pisoschi et al. 2018). In detail, we shall be reviewing in subsequent sections the natural or bio-based preservatives based on their source—plant-derived, animal-derived, and microbial-derived BIOBAMs and their possible mechanisms of action.

4.2.1 Plant-Based Preservatives

Natural and bio-based products from plant or plant extracts are known to contain some antimicrobial agents and could be employed as safe and environmentally friendly preservatives (Gyawali and Ibrahim 2014). Outlined below are some of the plant-based antimicrobial agents and their proposed mechanisms of action.

4.2.1.1 Plant Extracts

Plant extracts contain antimicrobial, antioxidant, and essential oils, which exhibit good preservation properties. Flavonoids, thiosulfinates, saponin, organic acids, phenolics, and glucosinolates can easily replace synthetic chemical preservatives

(Vilela et al. 2016). The phenolic compounds found in plants, such as acids, isoflavonoids, aliphatic alcohols, terpenes, and aldehydes, are the principal constituent with antimicrobial properties (Filip et al. 2021). These compounds are reducing agents which regulate hydrogen and suppress oxygen, leading to an antioxidant effect. The antimicrobial effects of grape and lemon fruit extract on food spoilage microorganisms *Photobacterium phosphoreum, Shewanella putrefaciens, and Pseudomonas fluorescens* have been reported in hamburgers with increased microbial stability (Corbo et al. 2008). Previous studies show that blackcurrant leaves and cherry tree phenolic extract exhibited good preservative antioxidant and antimicrobial activities in meat products increasing the shelf life of vacuum-packed sausages and significantly decreasing the level of malondialdehyde (MDA) produced in this product (Nowak et al. 2016). Previous studies show that leaves of green tea contain epigallocatechin, *epicatechin gallate*, Theaflavin monogallate A and B teaflavin gallate, epicatechin (Rohini et al. 2018). The antioxidant and antimicrobial activities of green tea against pathogens have been reported in previous literature (Demir 2021; Han et al. 2021; Kim et al. 2021; Sasagawa et al. 2021; Silva et al. 2021; Tarassoli et al. 2022; Zhang et al. 2021a, b). The antimicrobial activities of grape seed extract have also been reported (Al-Otibi et al. 2021; Hamida et al. 2021; Silvan et al. 2021). *S. typhimurium* and *E. coli* O157: H7 were decreased by grape seed extract and the development of *L. monocytogenes* and *Aeromonas hydrophila* in cooked meat (Aziz and Karboune 2017). Phenolic components of cranberry extract are the key antimicrobial factors in cranberry (Mousavi Khaneghah et al. 2018).

The exact mechanisms of plant extract action against spoilage microorganisms have not been elucidated. However, attacks on the phospholipid bilayers of the cell membrane, enzyme activity disruption, harming the microbe's genetic makeup, and oxidation of unsaturated fatty acids are some of the proposed pathways (Amiri et al. 2021). The inhibitory action of phenolic and aromatic compounds in plant extract against spoilage organisms results from an attack on the structure and function of the cytoplasmic membrane. Previous studies show that exposure to thymol and carvacrol dissipated the outer membrane of *S. typhimurium* and *E. coli* (Kachur and Suntres 2019). The denaturation of extracellular enzymes was proposed as the antimicrobial mechanism of nonphenolic isothiocyanates (Pisoschi et al. 2018). Terpenes exhibit their antimicrobial properties by disrupting the fat composition and the bacterial cell wall, consequently leading to the death of the bacteria (Aziz and Karboune 2017). Carvacrol acts by disrupting the outer membrane of Gram-negative bacteria, diffusing lipopolysaccharides, thereby improving the permeability of ATP, while for Gram-positive bacteria, its effects on the bacterial membrane affect the K^+ and H^+ cations (Tongnuanchan and Benjakul 2014).

4.2.1.2 Essential Oils

Essential oils are volatile natural compounds such as aldehydes, epoxides, terpenes, sulfides, ketones, esters, and amines determined by taste and odor (Ghamari et al. 2022; Mohajeri et al. 2021; Tongnuanchan and Benjakul 2014). They are mostly synthesized in the plastid and cytoplasm of plant cells via Methyl-D-erythritol-4-phosphate, malonic acid, and mevalonic acid pathways. Terpenes are hydrocarbons with several isoprene units, while terpenoids are produced through enzymatic alterations of terpenes via the transfer of methyl groups or the addition of oxygen molecules. Terpenes have been recorded to manifest significant antimicrobial properties (Pisoschi et al. 2018). The antimicrobial activities of coriander seed oil against bacteria, yeasts, molds, and dermatophytes have been reported (Silva and Domingues 2016). The essential oil derived from oregano was reported to exhibit an irreversible antimicrobial effect against *E. coli* (Pisoschi et al. 2018) due to their high phenolic compounds (y-terpinene, thymol ρ-cumene, and caracole) (Aziz and Karboune 2017). The antimicrobial actions of essential oils from parsley, lemongrass, rosemary, clove, oregano, garlic, coriander, bronze muscadine, cinnamon, and sage seeds against Gram-positive and Gram-negative bacteria have been reported (Aziz and Karboune 2017; Irkin and Esmer 2015). A previous study showed that lemon essential oils utilized as a microemulsion in salty sardine significantly reduced the microbial count of lactic acid bacteria (LAB), *Staphylococcus* spp., *Enterobacteria*, and low yield of histamine relative to the control (Alfonzo et al. 2017). Rosemary-derived essential oil showed antimicrobial effects against *Klebsiella pneumonia*, *S. aureus*, *E. coli,* and *Bacillus subtilis* (Aziz and Karboune 2017). The performance of essential oil can be enhanced by temperature, oxygen content, and low pH (Aziz and Karboune 2017; Dussault et al. 2014).

Essential oils from plants exert their antimicrobial activities through various mechanisms such as disrupting enzyme activities/structures, attacking the cell membrane, and compromising the formation of fatty acid hydroperoxide and bacterial genetic materials (Sohrabpour et al. 2021). The activity of phenolic compounds in essential oils is most likely dependent on changes in the permeability of bacterial cells, cytoplasmic membrane damage, interruption of the cellular ATP generation system, motility of protons, and cell death caused by cytoplasmic membrane damage (Mahmud et al. 2018). Furthermore, the presence and position of the hydroxyl group in the structure of phenolic compounds can be attributed to their antimicrobial properties.

4.2.1.3 Herbs and Spices

Herbs and spices are gotten from plant leaves and plant parts, respectively. Phenolic compounds such as flavonoids which are widely distributed in herbs, spices, and vegetables are metal ion chelators acting as catalysts in oxidation chemical reactions as well as inhibit cyclooxygenase and lipoxygenases, two key enzymes implicated in the progression of oxidative rancidity in food products (Embuscado 2015). The antimicrobial activities of some selected herbs have been widely studied (Abdul Qadir et al. 2017; De-Montijo-Prieto et al. 2021; Kamboj et al. 2021; Khan et al. 2021; Mahmud et al. 2018; Wang et al. 2015). Herbs are known to act against Gram-negative bacteria. The antimicrobial activities of cinnamon against *Salmonella anatum*, *S. aureus*, *L. monocytogenes*, *E. coli,* and *B. cereus* have been reported

(Shan et al. 2007). Similarly, the antimicrobial activities of different spices have also been reported (De-Montijo-Prieto et al. 2021; Mayekar et al. 2021; Meshaal et al. 2021; Ogwaro et al. 2021; Rahman et al. 2021; Vijayakumar et al. 2021).

4.2.2 Animal-Based Preservatives

There are diverse antimicrobial preservatives in animal sources, ranging from antimicrobial peptides, enzymes, bioactive polysaccharides, and lipids (Abdelhamid and El-Dougdoug 2020).

4.2.2.1 Enzymes

A few enzymes from animal sources have been reported to possess antimicrobial activities that hinder the growth of different food spoilage pathogens. The most common examples are the lysozymes and lactoperoxidase (LPO) (Singh 2018).

(a) ***Lysozymes and their mechanism of antimicrobial actions***

Lysozymes are potent bacteriolytic enzymes majorly sourced from egg white of chicken as well as milk, beef, and blood. Lysozyme has been used as antimicrobial preservatives for fish, meat, milk, and dairy products (Mei et al. 2019). Generally, studies have shown that lysozyme catalytically attacks the bacterial peptidoglycan by hydrolyzing the 1,4 glycosidic bonds among *N*-acetylmuramic acid and *N*-acetyl glucosamine, thereby destroying the structural architecture of the microbial cell wall (Singh 2018). The mode of action of lysozymes confers its effectiveness as an antimicrobial compound against Gram-positive bacteria (organisms with peptidoglycan rich cell walls) rather than Gram-negative organisms (having a lipopolysaccharide external layer). Nevertheless, a few findings have shown the effectiveness of lysozyme against Gram-negative bacteria when administered together with either surfactants or detergents and chelating agents. These detergents and chelating agents thwart the outer lipopolysaccharide layer of the Gram-negative bacteria, making them more susceptible to the lytic activities of lysozymes (Ferraboschi et al. 2021).

One interesting advantage of lysozyme, unlike other enzymes, is their enormous stability and resistance to denaturation by a wide range of pH and temperature, hence fostering their commerciality (Yang and Lesnierowski 2019). Inovapure is one of the most popular commercial brands for lysozyme and has been described in several studies as effective for various organisms such as *Listeria monocytogenes, Clostridium species, Bacillus* spp., *Pseudomonas* spp., *Salmonella* spp., and many others (Ferraboschi et al. 2021). Health Canada has authorized egg-derived lysozyme for preserving hard cheese from the gas-blowing activities of *C. tyrobutyricum* (Aziz and Karboune 2017).

(b) ***Lacto-peroxidase system and its mechanism of antimicrobial actions***

The lactoperoxidase system is another group of potent antimicrobial enzymes comprising lactoperoxidase, thiocyanate, and hydrogen peroxide. This enzyme system is commonly sourced from milk, tears, and saliva. Unlike lysozyme, which acts specific only on Gram-negative microbes, the lactoperoxidase system has broader applicability as it inhibits Gram-positive and Gram-negative bacteria and different fungi (Al-Baarri et al. 2019).

Generally, the mechanism of action of the lactoperoxidase system is as follows—using hydrogen peroxides, the lactoperoxidase enzymes catalyze the oxidation of thiocyanate to hypothiocyanite and its acidic derivatives (Zhang and Rhim 2022). Then, hypothiocyanite and its acids inhibit several microbial enzymes and proteins' sulfhydryl group, thereby resulting in microbial inactivation or death (Silva et al. 2020). Several studies have shown that the lactoperoxidase system is constitutively expressed in many animal sources but remains inactivated because of the limited concentration of either thiocyanate or hydrogen peroxide (Yousefi et al. 2022). Hence, the addition of a regulated thiocyanate and sometimes hydrogen peroxides activates the de novo lactoperoxidase system. Notably, the use of hydrogen peroxides for the activation of the lactoperoxidase system sometimes causes some toxicity worries. Different studies have shown the antimicrobial activity of the lactoperoxidase system (Zarei et al. 2010). Chiraz et al. (2013) showed the effectiveness of the bovine lactoperoxidase in inhibiting *Salmonella enteric Hadar,* the common foodborne causative agent of human gastroenteritis in Tunisia. The study also pointed out that the lactoperoxidase system can be subdued in the presence of starch around 100 g/L and above (Chiraz et al. 2013). Hence, the starch content of the food must be considered when adopting the lactoperoxidase system.

4.2.2.2 Antimicrobial Peptides

Many animals possess several short sequences of amino acids (peptides) as part of their defense mechanisms with inherent antimicrobial properties. This large group of bioactive peptides is commonly referred to as antimicrobial peptides (AMPs) and includes well-known defensins, pleurocidin, protamine, lactoferrin, magainin, and others (Liu et al. 2021). Many AMPs act on either or both Gram-negative and Gram-positive bacteria. Some even possess some antifungal, while others have shown antiviral properties. Generally, different AMPs may act differently (León Madrazo and Segura Campos 2020). Many AMPs may destabilize organisms' cellular lipid bilayer membrane and foster cell lysis while others either sequester vital components of the cells or inactivate important survival pathways. AMP has lately been reported to be an effective alternative for managing the menace of antibiotics resistance (Silveira et al. 2021). More details on some selected and common antimicrobial peptides from animal sources are discussed in the subsequent section.

(a) ***Pleurocidin***

Pleurocidin is a 25 amino acid peptide abundant in the skin of winter flounder (*Pleuronectes americanus*). Generally, pleurocidin has broad antimicrobial action against bacteria. The bioactive peptide destroys many organisms' lipid bilayer, causing leakage or even lysis of the cell (Erdem Büyükkiraz and Kesmen 2022). Due to several studies, pleurocidin has gained wide application as a food AMP due to its stable properties in high temperature, salt concentration, and wide pH range. Foodborne pathogens, including *Vibro parahemolyticus, E. coli, Listeria monocytogenes,* and many more, have been inhibited or killed by pleurocidin (Erdem Büyükkiraz and Kesmen 2022; Raju et al. 2020). Despite its advantage, one limitation of pleurocidin is its susceptibility and inactivation in divalent minerals such as calcium and magnesium (Wang et al. 2019). Hence, its application as a preservative in food rich in calcium or magnesium would not be effective.

(b) ***Defensins***

Defensins are a larger group of AMPs found on epithelial cells and host defense tissues of vertebrates such as chickens, turkeys, bovines, and others. Defensins also have broad antimicrobial action against bacteria (both Gram-positive and negative), fungi, and even envelop viruses (Amso and Hayouka 2019).

(c) ***Lactoferrin***

Lactoferrin is a glycopeptide that possesses broad-spectrum antimicrobial activities against Gram-positive and negative bacteria, parasites, and fungi (Erdem Büyükkiraz and Kesmen 2022). Lactoferrin elicits its antimicrobial activities through two possible mechanisms. First, due to its strong affinity for iron, lactoferrin sequesters most of the microbial iron, making them bio unavailable for vital enzymatic processes and consequently inactivating the cells. Secondly, lactoferrin having a large and distributed cationic patches on its surface destabilizes the anionic lipid A of lipopolysaccharide on Gram-negative bacteria surfaces, fostering membrane leakage (Pisoschi et al. 2018). Several studies have shown the effectiveness of lactoferrin against foodborne pathogens against *Pseudomonas* sp., *E. coli, Salmonella* sp., *Klebsiella* sp., *Carnobacterium* sp., and *L. monocytogenese* (Al-Nabulsi et al. 2009; Al-Nabulsi and Holley 2005; Colak et al. 2008; Juneja et al. 2012; Lönnerdal 2011).

Other animal-based AMPs are magainins from amphibians, casocidin, the hydrolytic product of S2-casein from milk, as well as diptericin, hymenoptaecin, Sarco toxin IIA, attacin, and coleoptericin from different insects (Guha et al. 2021; Józefiak and Engberg 2017; Mcmillan et al. 2020; Nath et al. 2021; Wu et al. 2018).

4.2.2.3 Bioactive Polysaccharides and Lipids

Some polysaccharides and lipids have been discovered to play several bioactive roles apart from their storage functions. A few polysaccharides such as chitosan and its derivatives, and lipids such as docosahexaenoic acid (DHA) and eicosapentaenoic acid (EPA), have been implicated with potent antimicrobial activities (Park et al. 2021)

(a) ***Chitosan and its derivatives***

Chitosan is the deacetylated derivative of chitin and one of the cationic polysaccharides comprising acetylglucosamine and glucosamine molecules, linked by β1–4 glycosidic bond. Natural sources of chitosan are shrimps shell, a few fungi, and green algae. Chitosan has been long known for its antimicrobial bioactivity against a broad spectrum of bacterial, yeast, and molds (Yu et al. 2021). However, its antimicrobial effectiveness is more prominent in Gram-negative bacteria than in other microbes due to its distortion of the anionic oligosaccharide of the extracellular membrane of bacteria (Zhang et al. 2021a, b). There are several routes or mechanisms by which chitosan performs its antimicrobial activities; they are as follows:

- Chitosan is cationic due to the positive charge on the second carbon of glucosamine, significantly below pH 6. The cationic properties confer improved solubility. It destabilizes the anionic lipopolysaccharides of the cellular membrane, thereby resulting in leakage of the cells' intracellular constituents and microbial death (Mesgari et al. 2021). One limitation of chitosan is its ineffectiveness as an antimicrobial membrane disruptor at neutral or alkaline pH, as its cationic properties are lost with increased pH (Aziz and Karboune 2017).
- Chitosan has been reported to act as a chelating agent, inhibiting microbial growth and toxin production by sequestering trace metals and vital microbial survival nutrients (Aziz and Karboune 2017).
- Chitosan has also been reported to inhibit valuable microbial enzymes and boost the host's immune response (Butnaru et al. 2019).
- Chitosan of smaller molecular weight can penetrate the cell to the nucleus of the microorganism, interfering with the organism's genetic regulation and expression pattern. Hence, lower molecular weight chitosan is more effective than higher weight counterparts, especially against Gram-negative microbes (Aziz and Karboune 2017; Zheng and Zhu 2003).
- Conversely, chitosan of higher molecular weight is more potent against Gram-positive organisms due to the possible likelihood of forming an exterior polymer membrane preventing the passage of nutrients into the cells. Hence, for Gram-positive organisms, the bactericidal effects increase with the increase in the molecular weight of chitosan (Tian and Liu 2020).

N-alkylated disaccharide chitosan is a derivative of chitosan with similar anti-microbial activities against foodborne pathogens (Yang et al. 2005). This derivative of chitosan was identified in the search for possible ways to minimize chitosan's limitation and pH dependency. The *n*-alkyl derivative confers a permanent cationic charge and properties to the bioactive polysaccharides without interfering with other toxicity mechanisms (Pisoschi et al. 2018). Studies have shown the effectiveness of the N-alkylated disaccharide derivative in inhibiting *E. coli* and *S. aureus* at an optimum pH of 7.0–7.5 (Yang et al. 2005).

(b) ***Fatty acids and lipids***

Several lipids of animal origin have been implicated with antimicrobial actions against various organisms—parasites, fungi, Gram-positive and negative bacteria, and others. For example, the lipid from milk was reported to inhibit a few Gram-positive bacteria such as *Staphylococcus aureus, Clostridium botulinum, Bacillus* spp., *Listeria monocytogenes*, and Gram-negative bacteria such as *E. coli, Pseudomonas aeruginosa,* and *Salmonella enteritis* against many fungi such as *Saccharomyces cerevisiae, Aspergillus niger,* and *Candida albicans* (Tiwari et al. 2009). Similarly, fatty acids commonly found in tissues of deep-sea fishes and shellfish were reported to inhibit four Gram-positive bacteria and seven Gram-negative foodborne pathogens (Shin et al. 2007). Generally, many antimicrobial lipids exert their activities through one of the following mechanisms of action: (a) membrane and cell wall disruption, thereby inactivating the cells, (b) restriction of intracellular replication, (c) restriction of intracellular target enzymes, and (d) lipids such as monoacylglycerols have been reported to make microorganisms more susceptible to heat by lowering their heat resistance capacity (Fischer 2020).

4.2.3 Microbial-Based Preservatives

Microbial-based antimicrobial preservatives involve the biopreservation of foods using microorganisms and their metabolites to enhance food safety and extend the shelf life. This method can be used in place of chemical treatments (Gálvez et al. 2007). The lactic acid bacteria (LAB) are the most commonly used microbe. Numerous LAB species are used as starter cultures for the fermentation of cheese and yogurt. The LAB is known to synthesize a variety of antibacterial compounds, including organic acids, ethanol, and bacteriocins. These compounds can inhibit or eliminate food spoilage microbes, and as a result, they are frequently used to preserve vegetables and meat products (Pisoschi et al. 2018). Both Gram-positive and Gram-negative bacteria produce bacteriocins. Bacteriocins are ribosomally produced proteins or peptides that, when released by the bacteria that produce them, can attach to the appropriate receptor on the surface of susceptible bacteria (often a similar or closely related bacterial strain) and kill or inhibit the bacteria (Yang et al. 2014). Bacteriocins are frequently employed in the biopreservation of food products because they do not alter the nutritional qualities of foods, are nontoxic, are potent at few concentrations, and always active during storage, making them an ideal biopreservation agent (Mahmud and Khan 2018; Milićević et al. 2021). *Lactobacillus, Carnobacterium, Lactococcus, Pediococcus, Streptococcus, Leuconostoc,* and *Propionibacterium* are major bacteria that commonly produce bacteriocin. Bacteriocins are classified into three: primary structure, biological activities post-translational modification, molecular weight, and biosynthetic mechanisms. Class I consists of antibiotics; class II consists of unmodified peptides; and class III consists of bigger heat-labile proteins (Du et al. 2022). Nisin, pediocin, enterocin, natamycin, and reuterin are the major bacteriocins used in food industries.

4.2.3.1 Nisin

Nisin is a type of bacteriocin that some *Lactococcus* species produce. It is usually considered safe and has been approved for use in the food sector (Silva et al. 2018; Surati 2020). Nisin has been shown to be antibacterial against Gram-positive bacteria such as *S. aureus* and *Listeria monocytogenes* and inhibit the offshoot of spores from various *Clostridium* and *Bacillus* species. It is, however, ineffective against Gram-negative bacteria, molds, and yeasts. Nisin acts through insertion and pore creation in the cytoplasmic membrane, resulting in the release of monovalent cations, amino acids, and ATP (Gyawali and Ibrahim 2014). Gram-negative bacteria are resistant to nisin's activity since their external layer restricts the passage of molecules larger than 700 Da; hence, nisin (3353 Da) cannot permeate the Gram-negative bacteria membrane. Additionally, nisin has been claimed to possess anti-biofilm capabilities (Siedler et al. 2019).

4.2.3.2 Pediocin

Pediocin is a cationic molecule with a low molecular weight. It contains 36–48 amino acid residues and comprises a hydrophilic N-terminal segment and a hydrophobic variable segment. It is predominantly produced by the *Pediococcus pentosaceus* and *Pediococcus acidilactici* (Porto et al. 2017). Pediocin inhibits the growth of various sensitive bacterial cells by acting on the cytoplasmic membrane during the development of spores. Pediocin inhibits the absorption of amino acids in the phospholipids of sensitive cells' cytoplasmic membranes. Additionally, it can produce pores in the cytoplasm membrane of the target cell (Kaur et al. 2013).

4.2.3.3 Enterocin

Enterococcus species produce enterococcal bacteriocins, which are frequently referred to as enterocins. They have received considerable research attention because of their activity against Gram-positive foodborne pathogens such as *Staphylococcus aureus*, *Bacillus cereus,* and *Listeria monocytogenes* (Khan et al. 2017). However, certain recently reported exceptions with broad activity spectra demonstrated the capacity to prevent the growth of Gram-negative microbes (De Kwaadsteniet et al. 2005; Line et al. 2008). Their spectra and modes of action differ, as do their molecular weights, molecular structures, thermal stabilities, genetic material, and pH range (Hwanhlem et al. 2017). Numerous enterocins have been shown to bind with specific bacterial cell receptors, including lipid II and other similar cell wall precursors, resulting in membrane permeabilization and target cell leakage. Additionally, enterocin NKR-5-3B has high affinity for negatively charged membranes, indicating that it can interact with negatively charged bacterial membranes of target cells and eventually kill them (Perez et al. 2021). Enterocin P is a cationic and amphipathic peptide that has been shown to associate with negatively charged bacterial membranes of target cells, leading to cell death (Herranz et al. 2001). The bulk of class III enterocins is significant, heat-labile proteins that can degrade sensitive cells by catalysis of cell wall breakdown. Enterolysin A can hydrolyze peptide bonds in target cells' peptidoglycan (Nilsen et al. 2003). Additionally, enterocins can kill bacteria by meddling with mRNA synthesis, transcription, and DNA replication (Cui et al. 2021).

4.2.3.4 Natamycin

Natamycin is a polyene macrolide antifungal produced by *Streptomyces* species. It is effective against practically all yeasts and molds but has minimal impact on bacteria, protozoa, and viruses (Pisoschi et al. 2018). Natamycin suppresses fungal development by irreversibly attaching to ergosterol, a key component of the fungal cell membrane. This affects the cell membrane's permeability, resulting in rapid leakage of critical ions and tiny peptides, resulting in cell lysis. Natamycin's numerous preservation benefits elucidate its attractive broad-spectrum antifungal biopreservative for foods and beverages. It is a nontoxic molecule that is effective at low doses and is active at even lower quantities. Natamycin is not resisted by fungi and is active over a broad pH range. Natamycin has no antagonistic effect on food quality, and due to its poor solubility, it does not migrate from the surface into the food, thereby preserving its organoleptic features. Additionally, natamycin can suppress rotting fungi without inhibiting the fermentation bacteria. It is chemically stable and has a long-lasting effect. Additionally, it is simple to apply and has a demonstrated track record of safety. Natamycin has been used as a biopreservative in food manufacturing for decades to suppress fungus growth in dairy products and other foods due to these properties.

4.2.3.5 Reuterin

The bacteriocin reuterin (*b*-hydroxypropionaldehyde) is produced by *L. reuteri*. It is water soluble and resistant to lipolytic and proteolytic enzymes. In anaerobic conditions, *L. reuteri* may convert glycerol to reuterin. Reuterin is an antibacterial agent with a broad spectrum of activity against foodborne spoilage microorganisms (Siedler et al. 2019). It has a significant function as a food biopreservative due to its water solubility, resistance to lipolytic and proteolytic enzyme destruction, and resist a wide range of temperature and pH conditions. Reuterin is effective against various common foodborne spoilage and pathogenic organisms, including *Enterococcus faecalis*, *Pseudomonas aeruginosa*, *Salmonella typhi*, *E. coli*, *S. aureus*, and *L. monocytogenes* (Arqués et al. 2011). However, the precise action mechanism of reuterin is unknown.

4.2.3.6 Organic Acids

Organic acids are present in all living organisms, with citric acid from citrus fruits and malic acid from apples being the most common. Acid levels in some immature fruits can be as high as 1%. These acids could play an essential function in protecting the fruit against pests and bacteria (Fan et al. 2018). Industrial production of organic acids from vegetables and fruits is prohibitively expensive, rendering them uneconomical. As a result, most organic acids are chemically manufactured or generated by fermentation. The oxidation of hydrocarbons and methanol carbonylation dominate industrial acetic acid production methods. Microorganisms produce nearly all citric acid, primarily through submerged fermentation of starch sucrose-based medium with *Aspergillus niger* (Yang et al. 2017). Microbial fermentation can also produce various organic acids, including lactic acids. Lactic acid bacteria, including *Leuconostoc, Pediococcus*, and *Lactobacilli,* are well studied to synthesize organic acids (Shi and Maktabdar 2022).

4.3 Recent Advances in Antimicrobial Preservatives: Bio-based Organic Acids Production

While physical, enzymatic, or microbiological extraction of bio-based antimicrobial preservatives may be appropriate, using genetically modified microorganisms (GMOs) as an antecedent for natural compounds is contentious. Although genetic modification techniques hardly alter natural compounds, most consumers believe that "natural" indicates "no GMOs." Citric acid, for example, is widely synthesized by *A. niger*. The fungus has been modified to boost citric acid production through genetic engineering. Because the antimicrobials produced by genetic engineering are not physically altered, it may be plausible to claim that they are natural (Fan et al. 2018).

However, chemicals are genetically modified from their native forms to improve their efficacy and durability in other cases. Some genes, for example, have been genetically engineered, and the consequent structures of molecules differ to some extent from their native forms. Also, plants are genetically engineered to create bacteriocins (through recombinant DNA). As a result, these chemicals produced through genetic engineering are bio-based antimicrobials (Rathod et al. 2021). Moreover, current advancement in metabolic engineering to produce industrially essential organic acids, such as lactic acid, succinic, and citric acid, emphasizes the utilization of high-performance microorganisms.

4.3.1 Lactic Acid

Lactic acid is biologically synthesized using lactic acid bacteria, but it is not economically convenient for industrial fermentations due to the need for moderate pH settings and nutritionally rich media (Okoye et al. 2022). Other natural producers, including *Saccharomyces cerevisiae,* have been investigated for lactic acid production due to their robustness, such as low pH tolerance and fewer dietary requirements. *S. cerevisiae* was able to grow by introducing the lactate dehydrogenase gene from *Pelodiscus sinensis* and modifying its expression. Rewiring cellular metabolic fluxes to produce lactic acid increased lactic acid production to 35 g/L by eliminating antagonistic pathways causing glycerol and ethanol synthesis. Furthermore, controlling internal redox availability by erasing external NADH dehydrogenase genes has proven significant for increased lactic acid synthesis, with the potential of synthesizing 117 g/L of lactic acid and constant pH of 3.5 (Chen and Nielsen 2016).

4.3.2 Succinic Acid

Succinic acid is primarily derived from petrochemicals and is mostly utilized as a chemical intermediary to synthesize refined chemicals, including perfume esters or food-grade neutralizing agents. Currently, microbial fermentation is being used on a much bigger scale, and several applications have begun the industrial production of succinic acid. The common bacterium *E. coli* has been modified metabolically to produce succinic acid. The removal of by-product generation and the TCA cycle's reductive route reinforcement resulted in a reasonably increased yield of succinic acid (Jansen and van Gulik 2014).

4.3.3 Citric Acid

Citric acid is an excellent bio-produced industrially organic acid type, steadily increasing application. Citric acid is primarily employed in the food and beverage sector, but it is also utilized in pharmaceuticals and various technical and industrial applications. *A. niger* is used in contemporary industrial production by overflown fermentation medium (Angumeenal and Venkappayya 2013). The utilization of yeast *Yarrowia lipolytica* has been reported to produce citric acid from various carbon sources, such as inulin, by exhibiting an exo-inulinase on its surface. Also, down-regulation of ATP-citrate lyase and up-regulation of isocitrate lyase resulted in increased citric acid synthesis (Liu et al. 2010), suggesting the potential of metabolic engineering in improving fermentation and yield of citric acid in *Y. lipolytica* (Soccol et al. 2006).

4.4 Conclusion

Consumer safety is one of the most serious considerations about traditional chemical preservatives in the food industry. Natural/bio-based compounds from microorganisms, animals, and plants have been investigated. The antimicrobial capabilities of these compounds have primarily been recommended for their safety as food preservatives. However, the challenge for researchers and industry is to realize their full potential by offering new techniques that will enable the large-scale industrial applications of these bio-based preservatives. Therefore, recent advances in antimicrobial preservatives have been uncovered for producing large quantities of organic acids from engineered microbes.

References

Abdelhamid AG, El-Dougdoug NK (2020) Controlling foodborne pathogens with natural antimicrobials by biological control and antivirulence strategies. Heliyon 6:e05020. https://doi.org/10.1016/J.HELIYON.2020.E05020

Abdul Qadir M, Shahzadi SK, Bashir A, Munir A, Shahzad S (2017) Evaluation of phenolic compounds and antioxidant and antimicrobial activities of some common herbs. Int J Anal Chem 2017:3475738. https://doi.org/10.1155/2017/3475738

Al-Baarri AN, Damayanti NT, Legowo AM, Tekiner IH, Hayakawa S (2019) Enhanced antibacterial activity of lactoperoxidase-thiocyanate-hydrogen peroxide system in reduced-lactose milk whey. Int J Food Sci 2019:8013402. https://doi.org/10.1155/2019/8013402

Alfonzo A, Martorana A, Guarrasi V, Barbera M, Gaglio R, Santulli A, Settanni L, Galati A, Moschetti G, Francesca N (2017) Effect of the lemon essential oils on the safety and sensory quality of salted sardines (Sardina pilchardus Walbaum 1792). Food Control 73:1265–1274. https://doi.org/10.1016/J.FOODCONT.2016.10.046

Al-Nabulsi AA, Holley RA (2005) Effect of bovine lactoferrin against Carnobacterium viridans. Food Microbiol 22:179–187. https://doi.org/10.1016/J.FM.2004.06.001

Al-Nabulsi AA, Osaili TM, Al-Holy MA, Shaker RR, Ayyash MM, Olaimat AN, Holley RA (2009) Influence of desiccation on the sensitivity of Cronobacter spp. to lactoferrin or nisin in broth and powdered infant formula. Int J Food Microbiol 136:221–226. https://doi.org/10.1016/J.IJFOODMICRO.2009.08.008

Al-Otibi F, Alkhudhair SK, Alharbi RI, Al-Askar AA, Aljowaie RM, Al-Shehri S (2021) The antimicrobial activities of silver nanoparticles from aqueous extract of grape seeds against pathogenic bacteria and fungi. Molecules 26:6081. https://doi.org/10.3390/MOLECULES26196081

Aluyor EO, Oboh IO (2014) PRESERVATIVES | traditional preservatives—vegetable oils. Encycl. Food Microbiol. 2nd ed, pp 137–140. https://doi.org/10.1016/B978-0-12-384,730-0.00263-9

Amiri S, Moghanjougi ZM, Bari MR, Khaneghah AM (2021) Natural protective agents and their applications as bio-preservatives in the food industry: an overview of current and future applications. Ital J Food Sci 33:55–68. https://doi.org/10.15586/IJFS.V33ISP1.2045

Amit SK, Uddin MM, Rahman R, Islam SMR, Khan MS (2017) A review on mechanisms and commercial aspects of food preservation and processing. Agric Food Secur 6:1–22. https://doi.org/10.1186/S40066-017-0130-8/TABLES/21

Amso Z, Hayouka Z (2019) Antimicrobial random peptide cocktails: a new approach to fight pathogenic bacteria. Chem Commun 55:2007–2014. https://doi.org/10.1039/C8CC09961H

Angumeenal AR, Venkappayya D (2013) An overview of citric acid production. LWT - Food Sci Technol 50:367–370. https://doi.org/10.1016/J.LWT.2012.05.016

Arqués JL, Rodríguez E, Nuñez M, Medina M (2011) Combined effect of reuterin and lactic acid bacteria bacteriocins on the inactivation of food-borne pathogens in milk. Food Control 22:457–461. https://doi.org/10.1016/J.FOODCONT.2010.09.027

Aziz M, Karboune S (2017) Natural antimicrobial/antioxidant agents in meat and poultry products as well as fruits and vegetables: a review. Meat Sci 58:486–511. https://doi.org/10.1080/10408398.2016.1194256

Bensid A, El Abed N, Houicher A, Regenstein JM, Özogul F (2020) Antioxidant and antimicrobial preservatives: properties, mechanism of action and applications in food—a review. Crit Rev Food Sci Nutr 62(11):2985–3001. https://doi.org/10.1080/10408398.2020.1862046

Butnaru E, Stoleru E, Brebu MA, Darie-Nita RN, Bargan A, Vasile C (2019) Chitosan-based bionanocomposite films prepared by emulsion technique for food preservation. Materials (Basel) 12(3):373. https://doi.org/10.3390/MA12030373

Chen Y, Nielsen J (2016) Biobased organic acids production by metabolically engineered microorganisms. Curr Opin Biotechnol 37:165–172. https://doi.org/10.1016/j.copbio.2015.11.004

Chiraz A, Ahlem M, Mariem H, Landoulsi A (2013) Optimizing antimicrobial activity of the bovine lactoperoxidase system against Salmonella enterica Hadar, a causative agent of human gastroenteritis in Tunisia. Afr J Microbiol Res 7:2719–2723. https://doi.org/10.5897/AJMR2013.5560

Colak H, Hampikyan H, Bingol EB, Aksu H (2008) The effect of nisin and bovine lactoferrin on the microbiological quality of Turkish-style meatball (tekirdağ köfte). J Food Saf 28:355–375. https://doi.org/10.1111/J.1745-4565.2008.00105.X

Corbo MR, Speranza B, Filippone A, Granatiero S, Conte A, Sinigaglia M, Del Nobile MA (2008) Study on the synergic effect of natural compounds on the microbial quality decay of packed fish hamburger. Int J Food Microbiol 127:261–267. https://doi.org/10.1016/J.IJFOODMICRO.2008.07.014

Cui Y, Luo L, Wang X, Lu Y, Yi Y, Shan Y, Liu B, Zhou Y, Xin L (2021) Mining, heterologous expression, purification, antibactericidal mechanism, and application of bacteriocins: a review. Compr Rev Food Sci Food Saf 20:863–899. https://doi.org/10.1111/1541-4337.12658

De Kwaadsteniet M, Todorov SD, Knoetze H, Dicks LMT (2005) Characterization of a 3944 Da bacteriocin, produced by Enterococcus mundtii ST15, with activity against Gram-positive and Gram-negative bacteria. Int J Food Microbiol 105:433–444. https://doi.org/10.1016/J.IJFOODMICRO.2005.03.021

Demir T (2021) Effects of green tea powder, pomegranate peel powder, epicatechin and punicalagin additives on antimicrobial, antioxidant potential and quality properties of raw meatballs. Molecules 26:4052. https://doi.org/10.3390/MOLECULES26134052

De-Montijo-Prieto S, del Carmen Razola-Díaz M, María Gómez-Caravaca A, Guerra-Hernandez EJ, Jiménez-Valera M, Garcia-Villanova B, Ruiz-Bravo A, Verardo V (2021) Essential oils from fruit and vegetables, aromatic herbs, and spices: composition, antioxidant, and antimicrobial activities. Biology (Basel) 10(11):1091. https://doi.org/10.3390/BIOLOGY10111091

Dey S, Nagababu BH (2022) Applications of food color and bio-preservatives in the food and its effect on the human health. Food Chem Adv 1:100019. https://doi.org/10.1016/J.FOCHA.2022.100019

Du R, Ping W, Ge J (2022) Purification, characterization and mechanism of action of enterocin HDX-2, a novel class IIa bacteriocin produced by Enterococcus faecium HDX-2. LWT 153:112451. https://doi.org/10.1016/J.LWT.2021.112451

Dussault D, Vu KD, Lacroix M (2014) In vitro evaluation of antimicrobial activities of various commercial essential oils, oleoresin and pure compounds against food pathogens and application in ham. Meat Sci 96:514–520. https://doi.org/10.1016/J.MEATSCI.2013.08.015

El-Saber Batiha G, Hussein DE, Algammal AM, George TT, Jeandet P, Al-Snafi AE, Tiwari A, Pagnossa JP, Lima CM, Thorat ND, Zahoor M, El-Esawi M, Dey A, Alghamdi S, Hetta HF, Cruz-Martins N (2021) Application of natural antimicrobials in food preservation: recent views. Food Control 126:108066. https://doi.org/10.1016/J.FOODCONT.2021.108066

Embuscado ME (2015) Spices and herbs: natural sources of antioxidants—a mini review. J Funct Foods 18:811–819. https://doi.org/10.1016/J.JFF.2015.03.005

Erdem Büyükkiraz M, Kesmen Z (2022) Antimicrobial peptides (AMPs): a promising class of antimicrobial compounds. J Appl Microbiol 132:1573–1596. https://doi.org/10.1111/JAM.15314

Fan X, Ngo H, Wu C (2018) Natural and bio-based antimicrobials: a review. ACS Symp Ser 1287: 1–24. https://doi.org/10.1021/bk-2018-1287.ch001

Ferraboschi P, Ciceri S, Grisenti P (2021) Applications of lysozyme, an innate immune defense factor, as an alternative antibiotic. Antibiotics 10:1534. https://doi.org/10.3390/ANTIBIOTICS10121534

Filip S, Đurović S, Blagojević S, Tomić A, Ranitović A, Gašić U, Tešić Ž, Zeković Z (2021) Chemical composition and antimicrobial activity of Osage orange (Maclura pomifera) leaf extracts. Arch Pharm (Weinheim) 354:2000195. https://doi.org/10.1002/ARDP.202000195

Fischer CL (2020) Antimicrobial activity of host-derived lipids. Antibiotics 9:75. https://doi.org/10.3390/ANTIBIOTICS9020075

Gálvez A, Abriouel H, López RL, Omar NB (2007) Bacteriocin-based strategies for food biopreservation. Int J Food Microbiol 120:51–70. https://doi.org/10.1016/J.IJFOODMICRO.2007.06.001

Ghamari MA, Amiri S, Rezazadeh-Bari M, Rezazad-Bari L (2022) Physical, mechanical, and antimicrobial properties of active edible film based on milk proteins incorporated with Nigella sativa essential oil. Polym Bull 79:1097–1117. https://doi.org/10.1007/S00289-021-03550-Y/FIGURES/6

Gokoglu N (2019) Novel natural food preservatives and applications in seafood preservation: a review. J Sci Food Agric 99:2068–2077. https://doi.org/10.1002/JSFA.9416

Guha S, Sharma H, Deshwal GK, Rao PS (2021) A comprehensive review on bioactive peptides derived from milk and milk products of minor dairy species. Food Prod Process Nutr 31(3): 1–21. https://doi.org/10.1186/S43014-020-00045-7

Gyawali R, Ibrahim SA (2014) Natural products as antimicrobial agents. Food Control 46:412–429. https://doi.org/10.1016/J.FOODCONT.2014.05.047

Hamida F, Syafriana V, Ramadhani CF, Nanda EV (2021) Antibacterial activity of grape seeds extracts (Vitis vinifera L.) against Streptococcus mutans ATCC 31987. J Farm Etam 1:50–58. https://doi.org/10.52841/JFE.V1I1.187

Han S, Abiko Y, Washio J, Luo Y, Zhang L, Takahashi N (2021) Green tea-derived epigallocatechin gallate inhibits acid production and promotes the aggregation of Streptococcus mutans and non-mutans Streptococci. Caries Res 55:205–214. https://doi.org/10.1159/000515814

Herranz C, Chen Y, Chung HJ, Cintas LM, Hernández PE, Montville TJ, Chikindas ML (2001) Enterocin P selectively dissipates the membrane potential of Enterococcus faecium T136. Appl Environ Microbiol 67:1689–1692. https://doi.org/10.1128/AEM.67.4.1689-1692.2001/ASSET/36B64646-E820-4DEF-97D1-E173D51DBCF8/ASSETS/GRAPHIC/AM0411585004.JPEG

Hwanhlem N, Ivanova T, Biscola V, Choiset Y, Haertlé T (2017) Bacteriocin producing Enterococcus faecalis isolated from chicken gastrointestinal tract originating from Phitsanulok, Thailand: isolation, screening, safety evaluation and probiotic properties. Food Control 78: 187–195. https://doi.org/10.1016/J.FOODCONT.2017.02.060

Irkin R, Esmer OK (2015) Novel food packaging systems with natural antimicrobial agents. J. Food Sci Technol 52(10):6095–6111. https://doi.org/10.1007/S13197-015-1780-9

Jansen MLA, van Gulik WM (2014) Towards large scale fermentative production of succinic acid. Curr Opin Biotechnol 30:190–197. https://doi.org/10.1016/J.COPBIO.2014.07.003

Józefiak A, Engberg RM (2017) Insect proteins as a potential source of antimicrobial peptides in livestock production. A review. J Anim Feed Sci 26:87–99. https://doi.org/10.22358/JAFS/69998/2017

Juneja VK, Dwivedi HP, Yan X (2012) Novel natural food antimicrobials. Annu Rev Food Sci Technol 3:381–403. https://doi.org/10.1146/ANNUREV-FOOD-022811-101241
Kachur K, Suntres Z (2019) The antibacterial properties of phenolic isomers, carvacrol and thymol. Crit Rev Food Sci Nutr 60:3042–3053. https://doi.org/10.1080/10408398.2019.1675585
Kamboj A, Jose S, Singh A (2021) Antimicrobial activity of natural dyes—a comprehensive review. J Nat Fibers. https://doi.org/10.1080/15440478.2021.1875378
Kaur G, Singh TP, Malik RK (2013) Antibacterial efficacy of Nisin, Pediocin 34 and Enterocin FH99 against Listeria monocytogenes and cross resistance of its bacteriocin resistant variants to common food preservatives. Braz J Microbiol 44:63–71. https://doi.org/10.1590/S1517-83822013005000025
Khan S, Ghosh S, Malik N, Bhat SA (2017) Antioxidant properties of garlic essential oil and its use as a natural preservative in processed food. Int J Chem Stud 5:813–821
Khan N, Jamila N, Amin F, Masood R, Atlas A, Khan W, Ain NU, Khan SN (2021) Quantification of macro, micro and trace elements, and antimicrobial activity of medicinal herbs and their products. Arab J Chem 14:103055. https://doi.org/10.1016/J.ARABJC.2021.103055
Kim JW, Kim CY, Kim JH, Jeong JS, Lim JO, Ko JW, Kim TW (2021) Prophylactic catechin-rich green tea extract treatment ameliorates pathogenic enterotoxic Escherichia coli-induced colitis. Pathogens 10:1573. https://doi.org/10.3390/PATHOGENS10121573
León Madrazo A, Segura Campos MR (2020) Review of antimicrobial peptides as promoters of food safety: limitations and possibilities within the food industry. J Food Saf 40:e12854. https://doi.org/10.1111/JFS.12854
Line JE, Svetoch EA, Eruslanov BV, Perelygin VV, Mitsevich EV, Mitsevich IP, Levchuk VP, Svetoch OE, Seal BS, Siragusa GR, Stern NJ (2008) Isolation and purification of enterocin E-760 with broad antimicrobial activity against Gram-positive and Gram-negative bacteria. Antimicrob Agents Chemother 52:1094–1100. https://doi.org/10.1128/AAC.01569-06/ASSET/B1B1859A-D4F6-446A-BC29-8E4A7B2B7449/ASSETS/GRAPHIC/ZAC0030871740001.JPEG
Liu XY, Chi Z, Liu GL, Wang F, Madzak C, Chi ZM (2010) Inulin hydrolysis and citric acid production from inulin using the surface-engineered Yarrowia lipolytica displaying inulinase. Metab Eng 12:469–476. https://doi.org/10.1016/J.YMBEN.2010.04.004
Liu Y, Sameen DE, Ahmed S, Dai J, Qin W (2021) Antimicrobial peptides and their application in food packaging. Trends Food Sci Technol 112:471–483. https://doi.org/10.1016/J.TIFS.2021.04.019
Lönnerdal B (2011) Biological effects of novel bovine milk fractions. Nestle Nutr Workshop Ser Pediatr Program 67:41–54. https://doi.org/10.1159/000325574
Mahmud J, Khan RA (2018) Characterization of natural antimicrobials in food system. Adv Microbiol 8:894–916. https://doi.org/10.4236/AIM.2018.811060
Mahmud J, Khan RA, Mahmud J, Khan RA (2018) Characterization of natural antimicrobials in food system. Adv Microbiol 8:894–916. https://doi.org/10.4236/AIM.2018.811060
Mayekar VM, Ali A, Alim H, Patel N (2021) A review: antimicrobial activity of the medicinal spice plants to cure human disease. Plant Sci Today 8:629–646. https://doi.org/10.14719/PST.2021.8.3.1152
Mcmillan KAM, Power Coombs MR, Bojarska J, Wolf WM, Remko M, Zielenkiewicz P, Saviano M, Zabrocki J, Kaczmarek K (2020) Review: examining the natural role of amphibian antimicrobial peptide magainin. Molecules 25:5436. https://doi.org/10.3390/MOLECULES25225436
Mei J, Ma X, Xie J (2019) Review on natural preservatives for extending fish shelf life. Foods 8: 490. https://doi.org/10.3390/FOODS8100490
Mesgari M, Aalami AH, Sahebkar A (2021) Antimicrobial activities of chitosan/titanium dioxide composites as a biological nanolayer for food preservation: a review. Int J Biol Macromol 176: 530–539. https://doi.org/10.1016/J.IJBIOMAC.2021.02.099

Meshaal AK, Hetta HF, Yahia R, Abualnaja KM, Mansour AT, Al-Kadmy IMS, Alghamdi S, Dablool AS, Emran TB, Sedky H, Batiha GES, El-Kazzaz W (2021) In vitro antimicrobial activity of medicinal plant extracts against some bacterial pathogens isolated from raw and processed meat. Life 11:1178. https://doi.org/10.3390/LIFE11111178

Milićević B, Tomović V, Danilović B, Savić D (2021) The influence of starter cultures on the lactic acid bacteria microbiota of Petrovac sausage. Ital J Food Sci 33:24–34. https://doi.org/10.15586/IJFS.V33I2.1918

Mohajeri N, Shotorbani PM, Basti AA, Khoshkhoo Z, Khanjari A (2021) An assessment of Cuminum cyminum (Boiss) essential oil, NaCl, bile salts and their combinations in probiotic yogurt. Ital J Food Sci 33:24–33. https://doi.org/10.15586/IJFS.V33ISP1.1990

Mousavi Khaneghah A, Hashemi SMB, Limbo S (2018) Antimicrobial agents and packaging systems in antimicrobial active food packaging: an overview of approaches and interactions. Food Bioprod Process 111:1–19. https://doi.org/10.1016/J.FBP.2018.05.001

Nath A, Csighy A, Eren BA, Nugraha DT, Pásztorné-Huszár K, Tóth A, Takács K, Szerdahelyi E, Kiskó G, Kovács Z, Koris A, Vatai G (2021) Bioactive peptides from liquid milk protein concentrate by sequential tryptic and microbial hydrolysis. Processes 9:1688. https://doi.org/10.3390/PR9101688/S1

Nilsen T, Nes IF, Holo H (2003) Enterolysin A, a cell wall-degrading bacteriocin from Enterococcus faecalis LMG 2333. Appl Environ Microbiol 69:2975–2984. https://doi.org/10.1128/AEM.69.5.2975-2984.2003/ASSET/0875C5E9-B9B5-4750-8F60-D97A77BFCCC5/ASSETS/GRAPHIC/AM053126106B.JPEG

Novais C, Molina AK, Abreu RMV, Santo-Buelga C, Ferreira ICFR, Pereira C, Barros L (2022) Natural food colorants and preservatives: a review, a demand, and a challenge. J Agric Food Chem 70:2789–2805. https://doi.org/10.1021/ACS.JAFC.1C07533/ASSET/IMAGES/ACS.JAFC.1C07533.SOCIAL.JPEG_V03

Nowak A, Czyzowska A, Efenberger M, Krala L (2016) Polyphenolic extracts of cherry (Prunus cerasus L.) and blackcurrant (Ribes nigrum L.) leaves as natural preservatives in meat products. Food Microbiol 59:142–149. https://doi.org/10.1016/J.FM.2016.06.004

Ogwaro BA, O'gara EA, Hill DJ, Gibson H (2021) A study of the antimicrobial activity of combined black pepper and cinnamon essential oils against Escherichia fergusonii in traditional African yoghurt. Foods 10:2847. https://doi.org/10.3390/FOODS10112847

Park J, Choi J, Kim DD, Lee S, Lee B, Lee Y, Kim S, Kwon S, Noh M, Lee MO, Le QV, Oh YK (2021) Bioactive lipids and their derivatives in biomedical applications. Biomol Ther (Seoul) 29:465. https://doi.org/10.4062/BIOMOLTHER.2021.107

Perez RH, Iwatani S, Zendo T (2021) Controlled functional expression of the circular bacteriocin enterocin nkr-5-3b and the leaderless bacteriocin lacticin Q. J Microbiol Biotechnol Food Sci 11:e4018. https://doi.org/10.15414/JMBFS.4018

Pisoschi AM, Pop A, Georgescu C, Turcuş V, Olah NK, Mathe E (2018) An overview of natural antimicrobials role in food. Eur J Med Chem 143:922–935. https://doi.org/10.1016/J.EJMECH.2017.11.095

Porto MCW, Kuniyoshi TM, Azevedo POS, Vitolo M, Oliveira RPS (2017) Pediococcus spp.: an important genus of lactic acid bacteria and pediocin producers. Biotechnol Adv 35:361–374. https://doi.org/10.1016/j.biotechadv.2017.03.004

Rahman MH, Asaduzzaman M, Kabir MS (2021) Determination of antimicrobial activity of traditional spices extracts against clinical isolates in Dhaka city. Stanford J Microbiol 11:17–19. https://doi.org/10.3329/SJM.V11I1.57147

Raju SV, Sarkar P, Kumar P, Arockiaraj J (2020) Piscidin, fish antimicrobial peptide: structure, classification, properties, mechanism, gene regulation and therapeutical importance. Int J Pept Res Ther 27:91–107. https://doi.org/10.1007/S10989-020-10068-W

Rathod NB, Phadke GG, Tabanelli G, Mane A, Ranveer RC, Pagarkar A, Ozogul F (2021) Recent advances in bio-preservatives impacts of lactic acid bacteria and their metabolites on aquatic food products. Food Biosci 44:101440. https://doi.org/10.1016/J.FBIO.2021.101440

Rohini S, Vishnupriya P, Rakshagan V, Jain AR (2018) Protective effects of theaflavin. Drug Invent Today 10:2097–2101

Sasagawa K, Domon H, Sakagami R, Hirayama S, Maekawa T, Isono T, Hiyoshi T, Tamura H, Takizawa F, Fukushima Y, Tabeta K, Terao Y (2021) Matcha green tea exhibits bactericidal activity against streptococcus pneumoniae and inhibits functional pneumolysin. Antibiotics 10: 1550. https://doi.org/10.3390/ANTIBIOTICS10121550

Okoye CO, Dong K, Wang Y, Gao L, Li X, Wu Y, Jiang J (2022) Comparative genomics reveals the organic acid biosynthesis metabolic pathways among five lactic acid bacterial species isolated from fermented vegetables. New Biotechnol. https://doi.org/10.1016/J.NBT.2022.05.001

Shan B, Cai YZ, Brooks JD, Corke H (2007) Antibacterial properties and major bioactive components of cinnamon stick (Cinnamomum burmannii): activity against foodborne pathogenic bacteria. J Agric Food Chem 55:5484–5490. https://doi.org/10.1021/JF070424D

Shi C, Maktabdar M (2022) Lactic acid bacteria as biopreservation against spoilage molds in dairy products—a review. Front Microbiol 12:4283. https://doi.org/10.3389/FMICB.2021.819684/BIBTEX

Shin SY, Bajpai VK, Kim HR, Kang SC (2007) Antibacterial activity of bioconverted eicosapentaenoic (EPA) and docosahexaenoic acid (DHA) against foodborne pathogenic bacteria. Int J Food Microbiol 113:233–236. https://doi.org/10.1016/j.ijfoodmicro.2006.05.020

Siedler S, Balti R, Neves AR (2019) Bioprotective mechanisms of lactic acid bacteria against fungal spoilage of food. Curr Opin Biotechnol 56:138–146. https://doi.org/10.1016/J.COPBIO.2018.11.015

Silva F, Domingues FC (2016) Antimicrobial activity of coriander oil and its effectiveness as food preservative. Crit Rev Food Sci Nutr 57:35–47. https://doi.org/10.1080/10408398.2013.847818

Silva CCG, Silva SPM, Ribeiro SC (2018) Application of bacteriocins and protective cultures in dairy food preservation. Front Microbiol 9:594. https://doi.org/10.3389/FMICB.2018.00594/BIBTEX

Silva E, Oliveira J, Silva Y, Urbano S, Sales D, Moraes E, Rangel A, Anaya K (2020) Lactoperoxidase system in the dairy industry: challenges and opportunities. Czech J Food Sci 38:337–346. https://doi.org/10.17221/103/2020-CJFS

Silva KA, Uekane TM, de Miranda JF, Ruiz LF, da Motta JCB, Silva CB, de Souza Pitangui N, Gonzalez AGM, Fernandes FF, Lima AR (2021) Kombucha beverage from non-conventional edible plant infusion and green tea: characterization, toxicity, antioxidant activities and antimicrobial properties. Biocatal Agric Biotechnol 34:102032. https://doi.org/10.1016/J.BCAB.2021.102032

Silvan JM, Gutierrez-Docio A, Guerrero-Hurtado E, Domingo-Serrano L, Blanco-Suarez A, Prodanov M, Alarcon-Cavero T, Martinez-Rodriguez AJ (2021) Pre-treatment with grape seed extract reduces inflammatory response and oxidative stress induced by Helicobacter pylori infection in human gastric epithelial cells. Antioxidants 10:943. https://doi.org/10.3390/ANTIOX10060943

Silveira RF, Roque-Borda CA, Vicente EF (2021) Antimicrobial peptides as a feed additive alternative to animal production, food safety and public health implications: an overview. Anim Nutr 7:896–904. https://doi.org/10.1016/J.ANINU.2021.01.004

Singh VP (2018) Recent approaches in food bio-preservation—a review. Open Vet J 8:104–111. https://doi.org/10.4314/ovj.v8i1.16

Soccol CR, Vandenberghe LPS, Rodrigues C, Pandey A (2006) New perspectives for citric acid production and application. Food Technol Biotechnol 44:141–149

Sohrabpour S, Rezazadeh Bari M, Alizadeh M, Amiri S (2021) Investigation of the rheological, microbial, and physicochemical properties of developed synbiotic yogurt containing Lactobacillus acidophilus LA-5, honey, and cinnamon extract. J Food Process Preserv 45:e15323. https://doi.org/10.1111/JFPP.15323

Surati S (2020) Bacteriocin, antimicrobial as a new natural food preservative: its potential and challenges. Erud Indones J Food Drug Saf 1:63–82. https://doi.org/10.54384/ERUDITIO.V1I1.34

Tarassoli Z, Najjar R, Amani A (2022) One-pot biosynthesis of silver nanoparticles using green tea plant extract/rosemary oil and investigation of their antibacterial activity. https://doi.org/10.1080/24701556.2021.2025086

Tian B, Liu Y (2020) Chitosan-based biomaterials: from discovery to food application. Polym Adv Technol 31:2408–2421. https://doi.org/10.1002/PAT.5010

Tiwari BK, Valdramidis VP, O'Donnell CP, Muthukumarappan K, Bourke P, Cullen PJ (2009) Application of natural antimicrobials for food preservation. J Agric Food Chem 57:5987–6000. https://doi.org/10.1021/JF900668N

Tomaska LD, Brooke-Taylor S (2014) Food additives: food additives—general. Encycl Food Saf 2: 449–454. https://doi.org/10.1016/B978-0-12-378612-8.00234-1

Tongnuanchan P, Benjakul S (2014) Essential oils: extraction, bioactivities, and their uses for food preservation. J Food Sci 79:R1231–R1249. https://doi.org/10.1111/1750-3841.12492

Torres-León C, Aguilar CN (2022) Food preservation. In: Quantitative methods and analytical techniques in food microbiology. Routledge, New York, pp 39–55. https://doi.org/10.1201/9781003277453-4

Vijayakumar G, Kesavan H, Kannan A, Arulanandam D, Kim JH, Kim KJ, Song HJ, Kim HJ, Rangarajulu SK (2021) Phytosynthesis of copper nanoparticles using extracts of spices and their antibacterial properties. Processes 9:1341. https://doi.org/10.3390/PR9081341

Vilela J, Martins D, Monteiro-Silva F, González-Aguilar G, de Almeida JMMM, Saraiva C (2016) Antimicrobial effect of essential oils of Laurus nobilis L. and Rosmarinus officinalis L. on shelf-life of minced "Maronesa" beef stored under different packaging conditions. Food Packag Shelf Life 8:71–80. https://doi.org/10.1016/J.FPSL.2016.04.002

Wang L, Yang R, Yuan B, Liu Y, Liu C (2015) The antiviral and antimicrobial activities of licorice, a widely-used Chinese herb. Acta Pharm Sin B 5:310–315. https://doi.org/10.1016/J.APSB.2015.05.005

Wang J, Dou X, Song J, Lyu Y, Zhu X, Xu L, Li W, Shan A (2019) Antimicrobial peptides: promising alternatives in the post feeding antibiotic era. Med Res Rev 39:831–859. https://doi.org/10.1002/MED.21542

Wu Q, Patočka J, Kuča K (2018) Insect antimicrobial peptides, a mini review. Toxins (Basel) 10(11):461. https://doi.org/10.3390/TOXINS10110461

Yang T, Lesnierowski G (2019) Changes in selected physicochemical properties of lysozyme modified with a new method using microwave field and oxidation. PLoS One 14:e0213021. https://doi.org/10.1371/JOURNAL.PONE.0213021

Yang TC, Chou CC, Li CF (2005) Antibacterial activity of N-alkylated disaccharide chitosan derivatives. Int J Food Microbiol 97:237–245. https://doi.org/10.1016/S0168-1605(03)00083-7

Yang SC, Lin CH, Sung CT, Fang JY (2014) Antibacterial activities of bacteriocins: application in foods and pharmaceuticals. Front Microbiol 5:241. https://doi.org/10.3389/FMICB.2014.00241

Yang L, Lübeck M, Lübeck PS (2017) Aspergillus as a versatile cell factory for organic acid production. Fungal Biol Rev 31:33–49. https://doi.org/10.1016/j.fbr.2016.11.001

Yousefi M, Nematollahi A, Shadnoush M, Mortazavian AM, Khorshidian N (2022) Antimicrobial activity of films and coatings containing lactoperoxidase system: a review. Front Nutr 9:229. https://doi.org/10.3389/FNUT.2022.828065/BIBTEX

Yu D, Yu Z, Zhao W, Regenstein JM, Xia W (2021) Advances in the application of chitosan as a sustainable bioactive material in food preservation. Crit Rev Food Sci Nutr 62(14):3782–3797. https://doi.org/10.1080/10408398.2020.1869920

Zarei M, Shahram Shekarforoush S, Khajehali E, Nazer AHK (2010) Antibacterial activity of lactoperoxidase system, activated by two different recommended methods against Escherichia coli O157:H7 in raw milk. Milchwissenschaft 65:298–301

Zhang W, Rhim JW (2022) Functional edible films/coatings integrated with lactoperoxidase and lysozyme and their application in food preservation. Food Control 133:108670. https://doi.org/10.1016/J.FOODCONT.2021.108670

Zhang Q, Zhang J, Zhang J, Xu D, Li Y, Liu Y, Zhang X, Zhang R, Wu Z, Weng P (2021a) Antimicrobial effect of tea polyphenols against foodborne pathogens: a review. J Food Prot 84:1801–1808. https://doi.org/10.4315/JFP-21-043

Zhang X, Ismail BB, Cheng H, Jin TZ, Qian M, Arabi SA, Liu D, Guo M (2021b) Emerging chitosan-essential oil films and coatings for food preservation—a review of advances and applications. Carbohydr Polym 273:118616. https://doi.org/10.1016/J.CARBPOL.2021.118616

Zheng LY, Zhu JF (2003) Study on antimicrobial activity of chitosan with different molecular weights. Carbohydr Polym 54:527–530. https://doi.org/10.1016/J.CARBPOL.2003.07.009

Chapter 5
Prebiotic and Synbiotic Foods

Uday S. Annapure and Harsh B. Jadhav

Abstract Human health primarily depends on the diet, as diet plays a fundamental role in revamping the population of intestinal microbiota, which assists in maintaining equanimity of various biological processes in the human body and also prevents diseases. Hence consumers are in demand for food that beneficially affects intestinal microbiota and improves human health. In this vein, prebiotic foods portray the most important foods which are known to perpetuate the load of beneficial microbiota and promote human health. Prebiotics are the fermentable fiber that fastidiously nourishes healthy bacteria and conserves beneficial microbiota habitat in the human gut. The beneficial effect of prebiotics can be enhanced and maintained for a longer time if they are taken in combination with probiotics. Probiotics are live microbes that improve flora in the gut and positively improve human health. The synergistic effect of both prebiotic and probiotic is synbiotic in which a population of live microbes (probiotics) is maintained/increased by using nondigestible fermentable fiber (prebiotics) and is an asset for human health. This chapter brings forth the inexorable background and important details on the sources, characteristics, potential health benefits, and future trends of prebiotic and synbiotic foods across the globe.

Keywords Prebiotic · Synbiotic · Probiotic · Nutraceutical · Health-friendly · Beneficial microbes

U. S. Annapure (✉)
Department of Food Engineering and Technology, Institute of Chemical Technology, Mumbai, India

Institute of Chemical Technology, Jalna, India
e-mail: us.annapure@ictmumbai.edu.in

H. B. Jadhav
Department of Food Engineering and Technology, Institute of Chemical Technology, Mumbai, India

A. K. Nadda, G. Goel (eds.), *Microbes for Natural Food Additives*, Microorganisms for Sustainability 38, https://doi.org/10.1007/978-981-19-5711-6_5

5.1 Introduction to Prebiotic (PB)

The gastrointestinal tract (GT) in the human body is considered a host for numerous microbes which are usually referred to as microbiota/microflora. With an increase in demand from consumers for healthy food, these microflorae are highly exploited under functional foods. The two main components which are considered to propitiously modify GT microflora are prebiotics and probiotics. "*A substrate which is selectively utilized by a host microorganism conferring health benefits is referred to as prebiotic*" (Hawrelak 2020). Prebiotics are generally promulgated to positively modify the GT microflora and are predicted to promote microflora of the epidermis, mouth, etc. As per the definition of prebiotics *host microorganism* refers to all the health-beneficial microbiota present in the entire GT which are benefited by the prebiotics (Scott et al. 2020). Prebiotics are nonviable components of food that are resistant to the process of digestion and absorption in the stomach rather they are easily fermentable by the health-friendly microflora present in the intestinal tract. The most prevailing prebiotics include lactulose, fructooligosaccharide (FOS), xylooligosaccharides (XOS), isomaltooligosaccharide (IMO), oligofructose (OF), galactooligosaccharides (GOS), rafinooligosaccharides (ROS), starch and nonstarch polysaccharides (Rosa et al. 2021). Generally the carbohydrate-based prebiotics have been dominating the prebiotic market but recently it is seen that components like carotenoids, polyphenols, herbs, polyunsaturated fatty acids, etc are also playing a key role. There are numerous polyphenols found in plant sources that reach the colon unaffected and are utilized by microorganisms efficiently thus showing a prebiotic effect (Singh et al. 2020). However, if a certain food constituent reaches the colon undigested but is not selectively utilized by the beneficial microflora but is consumed by a pathogenic microorganism then such food ingredients are not classified as prebiotics. On entry of prebiotics in the colon, they are favorably acted upon by the microflora having the ability to metabolize them; this positively favors the growth and activity of health-friendly microflora. Some of the food constituents reach the colon undigested but are not selectively acted upon by the microorganisms; such foods are known as colonic foods. Consumption of prebiotics is related to numerous positive physiological effects which include prevention of colon cancer, strengthening of the immune system, boosting bone health with improved absorption of minerals, preventing inflammation, increasing satiety and reducing the risk of obesity, decreasing total cholesterol, ameliorating lipid profile of blood and also maintaining heart health, antidiabetic (Pena 2007; Pokusaeva et al. 2011; Shafi et al. 2019; Zepeda-Hernández et al. 2021), and most importantly prebiotics help in improving and maintaining beneficial microflora of gut (Duque et al. 2021; Wang et al. 2012). Recently published literature has revealed that prebiotics are helpful in maintaining the respiratory tract by combating inhaled pollutants (Keulers et al. 2022) and are also beneficial in curing and preventing novel coronavirus infection (Bhagwat et al. 2020; Xavier-Santos et al. 2022a, b). Apart from this, prebiotics also find important application in positively affecting microflora of the urogenital tract in the female, epidermis, and oral cavity (Al-Ghazzewi and Tester 2014).

Over the last few decades prebiotics have attracted the attention of many consumers owing to their health-friendly characteristics. They are used as a functional ingredient in various food formulations and have occupied the processed food market in the form of beverages, cookies, biscuits, bread, fruit juices health bars, smoothies, extruded products, chocolates, yogurt, buttermilk, dahi, ice cream, soymilk, frozen desserts, jellies, infant foods, etc. (Ashwini et al. 2019; Bultosa and Dawa 2019; Mollakhalili-Meybodi et al. 2021). The addition of prebiotics in food not only improves the nutritional characteristics of food but also positively enhances the physicochemical, textural, and sensory attributes of processed food making it more consumer appealing. In a recent study by Coelho et al. (2019) lactulose was used in the formulation of fermented sausages. The authors reported that the prebiotic ingredient has proved to have a beneficial effect on textural, microbiological, and sensory attributes of sausages. Similarly, da Silva and Conti-Silva (2018) also reported in their study the beneficial effect of using FOS and inulin in the formulation of chocolate cookies. The authors reported that the prebiotic components positively affected the population of beneficial microbiota but also showed enhanced sensory attributes, more soluble, insoluble fibers, and higher fruit content. The health benefits and excellent characteristics imparted to food products by prebiotic ingredients have resulted in the huge market growth of prebiotics. The global market for prebiotic ingredients alone has jumped from 3.4 billion US dollars in 2018 to 6.3 billion US dollars in 2022 and is expected to cross the level of 8.34 billion US dollars by the end of 2026 (Cezar et al. 2017; Rosa et al. 2021).

5.2 Journey of Prebiotics

In India, the use of prebiotics in diet dates back to the ancient period around 4000 BC. The main diet of ancient India was based on cereals, pulses, and millets which contained dietary fibers that helped in easy digestion and positively benefited serum cholesterol levels. Legumes were also among the most consumed food crop in ancient India, and they contain a higher proportion of resistant starches which were beneficial as a prebiotic. Other than this Indian diet in ancient times also included milk and milk products, meat, honey, root crops (like garlic), fruits, and vegetables which were a storehouse of many prebiotics. In India, prebiotics were consumed in adequate quantity through daily diet (Samanta et al. 2011). Now, the concept of prebiotics has gained momentum and it is used as a functional ingredient in food across the globe. In the early 1950s Scientist Gyorgy found that *N*-acetyl-glucosamine promotes *Bifidobacterium* growth; soon after this in 1957 lactulose was also found to be beneficial for *Bifidobacterium* (Tymczyszyn et al. 2014), and this gave a new horizon to prebiotics. The first technical definition of prebiotics came in 1995 (Gibson and Roberfroid 1995), and from then the interest of food scientists has shifted towards prebiotics and many sources of prebiotics are being explored across the globe. Prebiotics have emerged as a novel pathway for modulating microflora of the gut. These components are selectively metabolized in the gut by the microflora already present in the human gut. Foods like garlic, oat, chicory, wheat, onion, barley, asparagus, and soybean are natural sources of prebiotics but their intake in Western diets is not adequate; however, they are consumed in good amounts in Indian diets. In Western countries, an efficient pathway to increase the intake of prebiotics is through fortification of this prebiotics in foods like yogurt, bread, ice creams, frozen dessert, biscuits/cookies, etc., which are frequently consumed (Walton et al. 2013). The journey of prebiotics started long back with the subjugation of gastrointestinal events and presently prebiotics can be found on plates of every individual irrespective of their age, country, and sex, as evident from research that they are capable of modulating mixed microflora ecosystem bestowing numerous health benefits to humans.

5.3 Sources of Prebiotics

5.3.1 Cereals

Cereal grains are among the most consumed and economical food sources across the world. They are also an excellent source of prebiotic ingredients responsible for maintaining a balance of beneficial microbiota in GT. Cereals like wheat, barley, rice, sorghum, oats, millets, etc. are known to be the source of numerous prebiotic candidates showing prebiotic characteristics and the majority of them are carbohydrates mostly oligosaccharides and polysaccharides. The nonstarch polysaccharides which are obtained from cereal grain contain active ingredients like cellulose polysaccharides, arabinoxylan, fructans, inulin, α-glucan, β-glucan, etc. and their content varies in different cereals. Cellulose is found in the cell wall of cereals, which upon hydrolysis produces oligosaccharides that show prebiotic characteristics (Kumar et al. 2012). Apart from cellulose cereals also contain noncellulose polysaccharide like α-glucan, β-glucan, and arabinoxylans in a good proportion which also exhibits prebiotic characteristics. β-glucan is present in high proportion in barley (3–6%) and oats (3–5%) which are found to be beneficial to human health as they are linked with the lowering of serum cholesterol levels and are also beneficial in the reduction of postprandial response related to blood glucose (Sima et al. 2018; Staneva et al. 2014). Arena et al. (2016) reported that the synergistic effect of β-glucan from oats and barley and probiotic bacteria shows a promising effect as an anti-inflammatory, immunostimulatory response and inhibits atherosclerosis. Research on β-glucan proved its beneficial effect on cancer and the health of the respiratory tract (Lam and Cheung 2013). Wheat also contains arabinoxylan, β-glucan, cellulose, and lignin which play a key role as dietary fiber (Gebruers et al. 2010). Among these, approximately 50 wt% of dietary fiber fraction is contributed by arabinoxylan; they are present in a higher proportion in aleurone and the bran segment. Their prebiotic potential is well documented in the literature and is governed by their molecular size.

A recent study by Paesani et al. (2020) showed that the arabinoxylan extracted from hard wheat species showed excellent prebiotic potential than the arabinoxylan extracted from soft wheat variety; however, both arabinoxylans from hard and soft wheat variety showed prebiotic potential, proving to be a good prebiotic candidate. Cereal grains like wheat, triticale, barley, rye, and spelt contain arabinogalactan peptides which are glycoproteins; approximately 90% of wheat flour molecular mass is covered by oligosaccharides of arabinogalactans (Fincher and Stone 1974). A recent study by Harris et al. (2020) reported the prebiotic potential of arabinogalactan peptides. The authors used fecal samples for in vitro fermentation by a genus of *Bifidobacterium* producing acetate and reported that the arabinogalactan peptides have the potential to be prebiotic; however, the synergistic activity of arabinogalactan peptides and arabinoxylan showed much better results as compared with the single prebiotic component.

Fructans are also among the prebiotics which is found in abundant amount in cereal grains. The fructans content of cereals is variable: rye contains about 5.6%, wheat contains about 0.7–2.4%, barley contains about 0.9–2.3%, and oat contains less than 1% fructans (Abdi and Joye 2021; Verspreet et al. 2015). There are three major types of fructans, viz. inulin, graminans, and phlein (Tungland 2018). Among these inulin-type fructans are widely studied and exploited as a prebiotic; there are clinical studies also available in the literature supporting the prebiotic potential of inulin-type fructans (Pasqualetti et al. 2014; Peshev and Van den Ende 2014; Roberfroid et al. 2019; Verspreet et al. 2015). Inulin is also a potential substrate for enzymatic synthesis of fructooligosaccharide (FOS) which also has excellent bifidogenic characteristics and shows smashing prebiotic properties. FOS is also found in cereals; durum wheat contains the highest percentage of FOS. Immature wheat contains the highest FOS as compared with the mature one; the content of FOS decreases as the grain matures and the highest content is seen in the milky phase (Casiraghi et al. 2011; Demirci et al. 2019).

Cereals contain starch which is metabolized and absorbed in the upper gastrointestinal tract, but a small fragment of starch makes its way to the lower gastrointestinal tract and it is referred to as “Resistant Starch” (RS) (Bojarczuk et al. 2022). The RS can be easily fermented by the bacteria present in a colon and serve as a potential prebiotic growth substrate for the beneficial microbiota present in the gut. RS has numerous health benefits like reducing chances of obesity, enhancing the metabolism of lipids, antidiabetic, maintaining gut microbiota, etc. (Bojarczuk et al. 2022; Wen et al. 2022). Cereals as the principal component in the diet contain an ample amount of prebiotic candidates which probably are the excellent substrate for the growth of beneficial microbiota in the human gut. Cereals can also be used as a carrier of probiotics in the gut through diet to enhance the beneficial effect of prebiotics (Singu et al. 2020b). These cereals are also used in the production of various processed foods like bread, cookies, biscuits, cakes, etc., and efforts should be made that these prebiotic ingredients are not destroyed or damaged during adverse processing conditions and storage of processed food.

5.3.2 Pulses

Dry grains obtained from leguminous crops are known as pulses; they are the principal food crops of developing economies such as India, Africa, and China. The crucial carbohydrates which account for almost 50% of total carbohydrates present in pulses are starches. Some starches exist as resistant starches which reach the large intestine and are utilized by the beneficial microbiota and impose health benefits on humans. Many research reported in the literature has shown beneficial effects on the consumption of pulses such as antidiabetic, low risk of cardiovascular diseases, antiobesity, and lowering of serum cholesterol (Ha et al. 2014; Padhi and Ramdath 2017; Sievenpiper et al. 2009). Among pulses, lupin seeds contain the highest proportion of α-galactosides as an oligosaccharide with prebiotic potential (Gullón et al. 2015). Oligosaccharides like galactosides, raffinose, stachyose, and ciceritol are present in chickpea; it is reported that these oligosaccharides are utilized by *bifidobacteria* as a fermentable source of carbon in vitro, which promotes the growth of *bifido* species along with other beneficial microbiota. Fernando et al. (2010) studied the prebiotic effect of chickpea on the population of beneficial microbiota in human adults in the age group of 18–65 years. Subjects were given a diet containing chickpeas for 3 weeks, and the bacterial population in the fecal samples of these subjects was examined. The authors reported that the chickpea diet positively altered the population of microflora in GIT with enhancement in the species of *bifidobacteria* and a decrease in the species of pathogenic bacteria like *Clostridium*. Chickpea containing oligosaccharides, and resistant starches have the potential to harmonize the microflora of the gut and act as a probable source of prebiotic candidates which can be extracted and added to processed food for health-friendly results. Similar to chickpea, pinto beans are also a rich source of prebiotic components (Finley et al. 2007). Sheflin et al. (2017) also proved the potential of the navy bean to improve the microflora of the human gut. Subjects were given 35 g of cooked pea beans powder every day for 28 days. The authors reported that the subjects fed with pea beans showed improved diversity of microflora in the gut as compared with the control. The improvement in the gut microflora population also reduced the risk of colorectal cancer in humans (Baxter et al. 2019). Consumption of pulses results in low insulin response due to lower glycemic index of pulses and does not increase blood glucose level; it is also a beneficial food to improve the lipid profile and lowers systolic blood pressure. These beneficial effects are because of the presence of prebiotic components like oligosaccharides, resistant starches, and other components like phytosterols, insoluble fibers, folic acid, and saponins in pulses. These components no doubt show their beneficial effect on younger individuals, but they are equally active when consumed by older individuals bestowing health benefits to the older population (Abeysekara et al. 2012). Recently published studies on animals have shown the positive effect of pulses consumption on the population of colonic microflora (Graf et al. 2019; Lutsiv et al. 2021). Resistant starches are naturally present in pulses and are known to be an excellent prebiotic; the resistant starch content in pulses can be altered by cooking operations. A comparison between cereals, pulses, and tubers showed that cooking operation significantly improves the resistant starch in pulses in comparison with cereals and tubers (Brummer et al. 2015; Yadav et al. 2010). After the process of cooking the starch present in pulses undergoes the process of retrogradation which increases the amount of resistant starches in pulses. There are recent studies available in the literature showing a positive effect of resistant starches on the gut microflora in in vitro studies. Eslinger et al. (2014) studied the effect of resistant starch in yellow pea on the gut microflora in rats. The authors reported that the yellow pea fed diet resulted in an increase in the population of *Bifidobacteria* and the weight gain in rats fed with yellow pea was less as compared with control. Similarly, Xu et al. (2020) reported the effect of RS3 from cooked lentil starch (A) and RS2 from uncooked lentil starch (B) on the microflora of male mice in a 6 weeks study. The authors reported that "A" resistant starches showed excellent benefits like enhanced antioxidant activity, decreasing blood glucose level, maintaining serum lipid, and most importantly improved the population of beneficial microflora in the gut. Pulses form the most economical source of prebiotics and dietary fibers in the diets of the population in developing nations. However, developed nations like the United States are still far from achieving a dietary fiber daily intake of 14 g/1000 kcal (Thompson and Brick 2016) which is known to have a beneficial effect on the human gut microflora population (McGinley et al. 2020). An alternative to this could be the consumption of a diet rich in pulses as pulses on an average contain 14–30 g of dietary fibers per 100 g, which is higher in beans (23–32 g) and lower in lentils (18–20 g) (Kadyan et al. 2022).

5.3.3 Fruits and Vegetables

Fruits and vegetables (FAV) contain many functional components that are beneficial in reducing the risk of chronic diseases in humans (Drewnowski et al. 2008). Regular intake of fruits and vegetables offers the benefit of prebiotics along with other components like minerals and vitamins. FAV contains nondigestive oligosaccharides like fructooligosaccharide, raffinose family oligosaccharide, stachyose, raffinose, nystose, sugar polyols, etc. One of the research conducted by Jovanovic-Malinovska et al. (2014) on 41 vegetables and 32 fruits showed the presence of numerous bioactive components in FAV having prebiotic potential. The authors reported the presence of individual fractions of oligosaccharides in FAV; the content of fructooligosaccharide (FOS) like 1-kestose was highest in raspberry, watermelon, and cherry at 0.32 ± 0.009, 0.29 ± 0.009, and 0.22 ± 0.011 g/100 g, respectively. Higher content of nystose was found in pears (0.32 ± 0.012 g), sour cherry (0.22 ± 0.009 g), fig (0.19 ± 0.015 g), and melon (0.19 ± 0.015 g). 1F-β-fructofuranosylnystose was present in lower concentrations in some fruits like watermelon, raspberry, pear, nectarine, and apple. Raffinose and stachyose were detected in apricot and melons in lower concentrations, i.e., in the range of 0.01–0.05 g per 100 g. Similarly, prebiotic components were also found in vegetables, and their concentration was higher in vegetables as compared with fruits.

Leek, garlic, mushroom, dandelion, and fennel contained higher amount of 1-kestose, and nystose was found in higher amounts in cabbage (0.82 ± 0.010 g), artichoke (0.73 ± 0.015 g), broccoli (0.72 ± 0.012 g), and garlic (0.52 ± 0.026 g). 1F-β-fructofuranosylnystose was found only in garlic, dandelion, chicory, and fennel. Onion, scallion, artichoke, broccoli, cabbage, garlic, kale, and dandelion were among the vegetables containing the highest percentage of fructooligosaccharide and rafinooligosaccharides. Similar results were also reported by Jovanovic-Malinovska et al. (2015) showing a higher concentration of fructooligosaccharide and rafinooligosaccharides present in garlic, scallion, onion, Jerusalem artichoke, spring garlic, blueberry, raspberry, watermelon, and nectarine extracted using ultrasound-assisted process. The concentration of the prebiotic components varies with different fruits and vegetables but almost all FAV contains one or the other prebiotic component. The waste by-products generated after the consumption of fruits include peels, seeds, and sometimes damaged fruits also contain functional components having prebiotic potential. Fruit peels are known to contain various valuable components like dietary fibers (DF), polyphenols, antioxidants, aroma compounds, etc. (Helkar et al. 2016). Some of them show positive effects on the population of gut microflora and can be considered prebiotics. For instance, dietary fibers are positively utilized by the *Lactobacillus* species and *Bifidobacterium* species for their growth and survival. Fruit and vegetable waste contains almost 60–90% of dietary fibers that are associated with numerous health benefits to humans (Garcia-amezquita et al. 2018). Apple pomace contains approximately 4.5–47.4 g/100 g DF which is higher than what is present in the apple (Skinner et al. 2018). Similarly waste from other fruits like dragon fruit, banana, citrus fruit, mango, pineapple, berries, etc. also contains prebiotic components (Pop and Suharoschi 2021). There is a huge amount of waste generated from vegetables that also contain soluble fibers, insoluble fibers, resistance starch, DF, FOS, GOS, ROS, etc. having prebiotic potential. The amount of prebiotics from fruits and vegetable waste varies and it largely depends on the method of extraction adopted. A recent study by Zahid et al. (2021) showed the prebiotic potential of waste from mango, banana, and apple fruit. The authors reported that the apple peel powder increased the population of *Bifidobacterium animalis* subsp. Lactis by 1.84 log CFU/mL. A similar effect was observed for banana and mango peel powder on the growth of *Lactobacillus, rhamnosus L. casei,* and *Bifidobacterium lactis.* The population of these beneficial bacteria showed rapid growth with only 2% of the peel powder of each apple, banana, and mango. The prebiotic components can be extracted from these FAV waste and can be converted into powder for easy handling and better utilization, which can be further used as a functional ingredient in the formulation of various processed foods such as biscuits, cookies, cakes, chocolates, ice creams, soft drinks, and fruit-based beverages (Table 5.1).

Table 5.1 Development of processed food using fruits and vegetable waste

Fruit/ vegetable	By-product	Processed food	Prebiotic component	(%) use	References
Mango	Peels	Chips	Dietary fiber	15%	Clarisa Zepeda-Ruiz et al. (2020)
Carrot	Pomace	Biscuits	Dietary fiber, pectin, hemicellulose	10%	Phebean et al. (2017)
Mango	Peels	Drink	Prebiotic oligosaccharide	5%	Anwaar et al. (2020)
Pomegranate	Seeds	Bread	Dietary fibers, lignin	5%	Gül and Şen (2017)
Grapes	Pomace	Chocolate	Prebiotic component	5%	Gizem et al. (2020)
Pineapple	Pomace	Sausages	Dietary fibers, prebiotic components	Less than 5%	Montalvo-González et al. (2018)
Apple/carrot	Pomace	Ice creams	Dietary fibers, lignin, hemicellulose	1–3%	Inés et al. (2020), Wang et al. (2020)
Olive	Pomace	Powdered food	Dietary fiber, hemicellulose	1%	Ribeiro (2020)
Pomegranate	Seeds	Noodles	Dietary fibers	10%	Koca et al. (2017)
Potato	Peels	Cakes	Dietary fibers	5%	Ben et al. (2017)
Carrot	Pomace	Donuts	Dietary fibers, lignin, cellulose	6–7%	Nouri et al. (2017)

5.3.4 Dairy Products

Humans are using milk as a food since ancient days, and it has served as an excellent source of nutrients for individuals of all age groups. Milk and milk products are among the largely consumed food products by the consumer of all age groups. Consumption of dairy products containing prebiotics is closely related to various health benefits such as boosting immunity, modulating microflora in the gut, antidiabetic, and maintaining serum blood cholesterol, which are majorly studied (Rosa et al. 2021). Prebiotic components like FOS, ROS, GOS, inulin, and galactans are used during the preparation of dairy products and dairy-based infant formulas to overcome metabolic disorders, enhance the absorption of minerals, and also to improve the microflora population in the gut (De Morais 2016). The supplementation of dairy products with prebiotics has gained momentum due to health concern in consumers to strengthen their immune system to survive the COVID-19 pandemic. Their addition to dairy products also improves physical, chemical, textural, and sensory characteristics. In dairy products like ice creams, prebiotics is used as a replacement for the fat without any compromise in taste. Balthazar et al. (2017) studied the effect of the replacement of fat by prebiotic components in the preparation of sheep milk ice cream. The authors reported use of prebiotic components instead of fat resulted in the formation of ice cream with white color, excellent overrun, and ice crystals with a smaller size which is ideally desired by the consumers. The physicochemical properties of prebiotic ice cream were similar to those of full fat ice creams. Yogurt is among the highly consumed dairy product across the world and the supplementation of yogurt with prebiotics increases the population of beneficial bacteria. Dairy products are not a good source of fibers; thus, the addition of prebiotics help in maintaining the population of beneficial bacteria. A study by Özer et al. (2005) showed the effect of inulin and lactulose on the growth of *L. acidophilus* and *B. bifidum* in yogurt and its comparison with the yogurt without prebiotic components. The authors reported that the yogurt containing prebiotic components showed growth of *B. bifidum* bacteria and *L. acidophilus*, which was not observed in yogurt without prebiotic components. The prebiotic components are also found to increase the consistency of yogurts, improve texture, firmness, and hardness, increase creaminess, and obstruct the process of post acidification in yogurts (Balthazar et al. 2016; Costa et al. 2015). Like yogurts and ice creams, these prebiotic components are also being added to other dairy products like cheese (Speranza et al. 2018), kefir (Delgado-Fernández et al. 2019), dairy-based beverages (Souza et al. 2019), and milk chocolate (Valencia et al. 2016). It is also observed that the addition of these prebiotic components in dairy products may sometime have a negative impact on the physicochemical characteristics of dairy products (Delgado-Fernández et al. 2020). The supplementation of dairy products with prebiotics contributes to the enhancement of physicochemical, textural, sensory, and microbiological characteristics of dairy products when added in the optimum quantity. Addition below the optimum quantity may not result in the growth and development of probiotic bacteria, and addition above the optimum quality results in decreased textural and sensory attributes of dairy products. The quantum of research focusing on the incorporation of the prebiotic component in the dairy product has increased in the last few years because dairy products are an applaudable matrix for the addition of health-friendly components like prebiotics.

5.3.5 *Meat Products*

Meat and meat products have played a key role in the human diet for ages. It is a nutrient-dense food that serves as a source of fat-soluble vitamins, essential fatty acids, essential amino acids, high-quality proteins, minerals, etc. (Yousefi et al. 2018). Despite meat and meat products being the storehouse of nutrients, there are many studies reported in the literature showing a close association between the consumption of meat with various health-related issues (Domingo and Nadal 2017; Lippi et al. 2016). Given this, consumers are in demand for healthy nutrition-rich meat products (El Zeny et al. 2019; Illippangama et al. 2022). The demand from consumers forced food scientists to develop a healthier form of meat product by incorporating health-promoting ingredients that will reduce the possible baleful effects of consumption of meat products on human health (Ursachi et al. 2020). The incorporation of prebiotic ingredients into meat products has gained acceleration due to its health-friendly characteristics and positive effect on the qualities of meat products without affecting sensory attributes. Most of the prebiotic ingredients like

Table 5.2 Incorporation of prebiotic in meat and their effect of beneficial microflora

Meat product	Prebiotic	Concentration (%)	Conclusion	References
Sausages	Inulin	–	Increase in short-chain fatty acid and bifidobacteria species in fecal and plasma of rats fed with inulin containing sausages	Thøgersen et al. (2019)
Sausages	Inulin	16	An in vitro study showed that during storage there was increase in the population of lactic acid bacteria which increased to 10.09 log CFU/g after 14 days of storage	Glisic et al. (2019)
Salami	Inulin	2	In vitro study showed increase in the population of *Lactobacillus species* and decrease in the population of *Escherichia* spp. *and Shigella* spp.	Pérez-Burillo et al. (2019)
Italian type salami	FOS	2	The in vitro study showed an increase in the number of *Lactobacillus* spp. by 8 log CFU/g after 18 days of storage	Bis-Souza et al. (2020a)
Salchichon	FOS	2 g/100 g	Incorporation of FOS increased the viability of *Lactobacillus* spp. and decreased the population of *Enterobacteriaceae* in Spanish Salchichon	Bis-Souza et al. (2020b)

fructans, and inulin also act as dietary fiber which is beneficial in decreasing the risk of cancers, obesity, and heart-related diseases (Hathwar et al. 2012). Apart from the health-promoting effects of prebiotics, they also increase juiciness in meat products, enhance cooking yield, and also have a beneficial effect on the textural properties of meat (Talukder 2015). The major flaw associated with meat products is that it lacks dietary fibers, so different approaches are being undertaken to develop healthier meat products with the incorporation of prebiotics in it. Inulin, a well-known prebiotic component, is well explored in the meat industry as a fat replacer. Inulin is an oligosaccharide containing a long polyfructose chain that imparts a smooth mouthfeel and increases juiciness and creaminess in meat products owing to their property of forming a three-dimensional gel in a water-rich environment. Some of the studies have shown the positive effect of prebiotics added to meat products on the population of gut microflora (Table 5.2). Thogersen et al. (2019) studied the effect of inulin on the gut microflora in rats fed with meat sausages containing inulin. The authors reported that the rats fed with the sausages containing inulin showed a higher concentration of short-chain fatty acids present in the plasma metabolome and the fecal of rats. There was a significant increase in the population of *Bifidobacteria* species in the fecal of rats fed with sausages containing inulin as compared with the control group. Like inulin other prebiotic ingredients also showed a beneficial effect on the growth of probiotic bacteria. Bis-Souza et al. (2020a) investigated the effect of the addition of fructooligosaccharide in meat products on probiotic bacteria. The authors reported a significant increase in the population of *Lactobacillus* spp. and a decrease in the population of yeast and *Enterobacteriaceae* because of the formation of short-chain fatty acids which resulted in lowering the pH making it unsuitable for the survival of pathogenic bacteria. The addition of prebiotics to meat and meat products is an important source of prebiotics for humans taking a diet rich in meat products. In addition to functional ingredients like oligosaccharides (prebiotics), there are certain prebiotic peptides present in the meat. The hydrolysate of proteins from meat muscles is found to be beneficial for the growth of *Bifidobacteria* and lactic acid bacteria (Arihara and Ohata 2011). A skeletal muscle protein hydrolysate from pigs showed an increase in the growth of *Bifidobacteria.* The tripeptide extracted and purified showed the benefits of prebiotics and helped increase the population of *Bifidobacterium bifidum.* The dipeptide and the free amino acid which were part of this tripeptide did not affect the growth of *Bifidobacterium bifidum* (Arihara and Ohata 2011). There are chances of formation of meat peptides from meat protein during processes like the fermentation of meat, digestion of meat, ripening of meat, and curing of meat but the sequence of amino acids in these peptides plays an important role to be considered as a prebiotic. Meat and meat products are being continuously researched for being carriers of prebiotics in the diet of individuals across the globe. But, still there is huge scope for research in this area as many of the prebiotic ingredients still need to be checked for their potential benefits in meat products. The demand for such prebiotic meat products remains an unanswered query because consumers consider meat and related products as bad for their health because of higher saturated fat leading to health issues. With the increase in the scientific evidence related to prebiotic meat products, a consumer should be made aware of their health benefits; this will help in the expansion of the prebiotic meat product market globally.

5.3.6 Miscellaneous Sources

Minor sources like plant milk, tree nuts, and microbial sources are now emerging as a source of prebiotics. The plant milk/vegetable milk emerged as a replacement for dairy milk for consumers intolerant to dairy products. It includes cashew milk, hemp milk, pistachio milk, hazelnut milk, sunflower milk, walnut milk, corn milk, coconut milk, oat milk, almond milk, peanut milk, soy milk, etc. These are obtained from legumes, cereals, pulses, nuts, and seeds which in a raw state are a good source of prebiotics as discussed in previous sections. The simple procedure of obtaining plant milk is the treatment of raw materials like washing, cleaning, blanching, etc. The treated raw material is subjected to the pressing operation to extract the milk from it, which then undergoes purification or simple filtration to separate any foreign material and then stored till consumed (Kehinde et al. 2020; Vallath et al. 2022). The prebiotic components which were originally present in the raw material (namely cereals, nuts, pulses, legumes) are extracted in the plant milk and this can be consumed as it is or it can be added to other products to increase its palatability. The plant-based milk as a source of prebiotics and its uses in food products is an emerging field with a lot of research needed to check the potential of prebiotics in enhancing the growth of gut microflora. One recent study by Nissen et al. (2020) focused on the prebiotic potential of hemp seed milk fermented with *Lactobacillus* spp. The author reported that the growth of *Lactobacillus plantarum* and *Bifidobacterium bifidum* was higher in the hemp seed drink as compared with the control and also the prebiotic score as determined by the qPCR test showed a higher score for the hemp seed drink, showing that the drink contained prebiotics which were utilized by the beneficial bacteria for their survival and growth. The nuts which are considered a rich source of health-beneficial functional components are also a good source of prebiotics. Research shows that the skin and the outer covering of these nuts also contain some functional components having prebiotic potential (Surek and Buyukkileci 2017; Tunçil 2020). Almond shell contains xylooligosaccharides which is a prebiotic component and has a positive effect on the growth of beneficial microflora. Studies on almond skin showed that it positively increased the number of *Bifidobacteria* and decreased the growth of pathogenic bacteria. The DF present in the skin of almonds showed prebiotic potential and the consumption of powder generated from drying almond skin showed an increase in the *Bifidobacteria* and *Lactobacillus species* (Liu et al. 2014; Mandalari et al. 2010). Like almonds, there are research pieces of evidence of the presence of prebiotic ingredients in walnut (Ding and Li 2021; Holscher et al. 2018), cashew nuts (Sisconeto Bisinotto et al. 2021), hazelnut (Surek and Buyukkileci 2017), pistachio (Hesam et al. 2021), etc. The prebiotics will be effectively utilized by the probiotic bacteria present in the human gut if nuts are consumed as a whole, or nut milk can be extracted and then added to products like yogurt, biscuits, etc. or to aid the process of formulation of processed foods, the nut milk can be dehydrated and the powdered nut milk can be added to numerous processed food to utilize its health benefits, especially as a source of prebiotics (Udayarajan et al. 2022).

Marine algae are consumed by humans as a source of dietary fibers and bioactive health-friendly components and they have attracted consumers as a low-calorie food (Paiva et al. 2014). Marine algae are not new to processing industries; they are largely exploited in food, pharma, cosmetic, and fertilizer industries. The marine algae show prebiotic potential due to the presence of dietary fibers in it; this DF differs from the DF found in fruits and vegetables in quantity and also in technological aspects; hence the fermentation pattern of marine algae DF is different (Gupta and Abu-Ghannam 2011; Wells et al. 2017). The brown algae (BA), red algae (RA), and green algae (GA) are being used for consumption in countries like Korea, Japan, and China and now they have made their way onto the plates of consumers from other parts of the globe as well as owing to their prebiotic/health-friendly characteristics (Gómez-Ordóñez et al. 2010; Lange et al. 2015). BA contains alginate, fucoidan, and β-glucan as a potential source of DF which are not being metabolized by the enzymes present in the human body; thus, they are resistant to digestion by humans. Similarly, RA contains carrageenan, agar, glucan, gelidiales, and gracilariales whereas GA contains ulvans, xylose, mannose, galactose, and arabinose. These components found in algae are known to be effective against tumors (Menshova et al. 2014), show strong antioxidant potential (Fleita et al. 2015), and lower cholesterol (Qi et al. 2012), anti-inflammatory (Wijesinghe and Jeon 2012). Though marine algae have numerous health-promoting characteristics, the scientific evidence regarding DF from marine algae remains scarce. There is huge scope for future research in the field of marine algae and its utilization in food products and the study of the prebiotic potential of DF from algae.

5.4 Potential Health Benefits of Consuming Prebiotic Foods

Numerous species of microorganisms are inhabitant of the human intestine that lives in a mutualistic relationship with the host. Prebiotics are known to improve the survival of beneficial microflora in the human gut and these beneficial bacteria inhibit the growth of pathogenic microorganisms (Rosa et al. 2021). This proliferation of health-friendly microflora improves the mucosal membranes and helps in better absorption of minerals, vitamins, and other nutrients, reducing the risk of cancer, obesity, and diabetes (Table 5.3) (Colantonio et al. 2020; Hawrelak 2020).

5.4.1 Generation of Metabolites and Alteration of the Architecture of the Colon

Lactobacillus and *Bifidobacteria* spp. inhabited in the human gut are responsible for the fermentation of prebiotic components, thereby producing end products like acetate, propionate, and butyrate which represent about 95% of short-chain fatty acids present in the gut region (Bultosa and Dawa 2019). Butyrate is the principal energy substrate for the intestinal epithelium. Butyrate also has the ability to regulate the process of growth and differentiation of cells. It aids in the process of detoxification of cells and also reduces the possibility of cancer by programmed cell death. The other short-chain fatty acid like propionate is efficiently employed in the liver for the process of gluconeogenesis and acetate is metabolized and serves as a source of nutrition for muscles (Hawrelak 2020). The formation of short-chain fatty acid decreases the pH of the gut which directly reduces the proliferation of pathogenic microorganisms and helps in maintaining the population of beneficial microflora in the human gut (Blaut 2002).

The studies reported in the literature showed that prebiotics help in modifying the surface of the intestinal wall. Mineo et al. (2006) reported in their study that the effect of feeding rats with prebiotics resulted in higher calcium absorption and holding with improvement in the intestinal crypts in terms of depth, length, and width. The authors further reported that the prebiotic diet increased the quantity and strength of epithelial cells. The prebiotic also positively affected the absorptive portion of the cecum which was remarkably increased (Raschka and Daniel 2005). These positive changes in gut physiology were not observed in the axenic animals supplied with a diet containing prebiotics. This indicates the need for probiotic bacteria to be present in the gut which

Table 5.3 Health benefits of prebiotics

Prebiotic component	Human/ animal study	Investigation	Conclusion	References
Xylooligosaccharides	Animal	Study aimed to investigate the effect of consumption of xylooligosaccharides on accumulation of fat in mice	There was an increase in the level of short-chain fatty acid in feces, decrease in serum leptin concentration, decrease in perirenal and epididymal fat in mice	Long et al. (2019)
Prebiotic combination including fructooligosaccharide, xylooligosaccharides, resistance starch, and polydextrose	Human	Investigate the effect of prebiotic intake on the population of microbiota and immune function in colorectal patients before and after operation	Before operation intake of prebiotics increased level of transferrin. Immunoglobin G and A	Xie et al. (2019)
			After operation there was an increase in suppressor T cells, B lymphocytes, immunoglobin G and A	
Inulin	Human	Investigate the effect of consumption of prebiotics on concentration of short-chain fatty acid in person with obesity	The concentration of short-chain fatty acid increased after inulin intake, decrease in plasma free fatty acid, decrease in blood glucose	Van der Beek et al. (2019)
Dietary fiber	Human	To investigate the effect of dietary fiber on the level of insulin in humans with obesity	Decreased hyperinsulinemia	Provost et al. (2019)
Mannitol	Animal	To investigate the influence of prebiotic intake on activity of neocortical epileptiform	Decrease in neocortical epileptiform activity, decrease in neuronal volume	Glykys et al. (2019)
Fructooligosaccharide	Human	Systematic review focusing on the effect of fructooligosaccharide on mineral absorptions in humans	Absorption of calcium, iron, and magnesium was improved after intake of prebiotic	Costa et al. (2020)
Asparagus cochinchinensis neutral polysaccharide	Human	To investigate the effect of prebiotic extracted from	The concentration of acetate, n-valeric, i-valeric acid	Sun et al. (2020)

(continued)

Table 5.3 (continued)

Prebiotic component	Human/ animal study	Investigation	Conclusion	References
		Asparagus cochinchinensis on gut microflora	increased with lowering of pH in the gut. There was an increase in population of *Bifidobacterium, Megamonas*, and *Prevotella* with a decrease in population of *Haemophilus*	
Gun Arabic	Animal	To investigate the effect of prebiotics on diabetes	The authors concluded that the intake of prebiotic improved plasma lipid profile, maintained blood glucose level, and is effective against diabetes and kidney related disease	Khalil et al. (2021)
Inulin	Human	Effect of inulin on gut microflora in hemodialysis patients	The study has indicated that there was no significant increase in the microflora of gut and there is a need for further studies to investigate this effect on hemodialysis patients	Biruete et al. (2021)
Combination of xylooligosaccharides, fructooligosaccharide, and dextrin	Animal	To investigate the effect of prebiotic combination on inflammatory bowel diseases	Decrease in plasma IL-6, improved gut microflora, suppression of plasma pro-inflammatory agents. The combination was helpful in curing inflammatory bowel disease	Wong et al. (2022)
Resistant starches from raw banana	Human	To investigate the effect of prebiotics on the kidney transplant recipient	Prebiotic intake decreased the gastrointestinal symptoms in recipient and also moderate tolerability and adherence were observed in prebiotic group	Chan et al. (2022)

(continued)

Table 5.3 (continued)

Prebiotic component	Human/ animal study	Investigation	Conclusion	References
Ganoderma lucidum oligosaccharide	Human	To study the effect of new prebiotic on the gut microbiota in human feces	The prebiotic increased the concentration of short-chain fatty acids as checked in in vitro fermentation and increased the population of beneficial microorganism in human fecal matter	Lan Yang et al. (2022)
25% xylooligosaccharides and 75% fructooligosaccharide	Animal	To investigate the effect of prebiotics in lipid and glucose metabolism	The combination of prebiotics was effective against obesity, improved liver health, improved serum glucose level, and decreased oxidative stress in obese mice	Li et al. (2022)

effectively metabolized prebiotics and produces end products that help in the modification of the architecture of the gut (Blottiere et al. 2003).

5.4.2 Improved Absorption of Minerals

Prebiotics are known to help in the retention of minerals and enhance their absorption in the human body, thereby overcoming the problem of osteoporosis and repairing fractures in bones. The documented research shows the beneficial effect of consumption of prebiotics on retention and absorption of minerals specifically magnesium and calcium (Beynen et al. 2002). The principal site for the absorption of calcium is the distal intestine. The rate of absorption of minerals is governed by the rate of fermentation of prebiotics by the probiotic cells present in the intestine (Kumari et al. 2020). Due to the formation of short-chain fatty acid, there is a drop in the pH of the colonic region, which enhances the solubility of calcium and improves the expression of the calcium transport pathway (calbindin) (Scholz-Ahrens et al. 2008). There are both animal studies (Ohta et al. 2000; Younes et al. 2001) and human studies (Trinidad et al. 1996) available in the literature showing the close relation between consumption of prebiotics and enhanced calcium absorption. Abrams et al. (2005) studied the effect of consumption of a diet rich in inulin-type prebiotics on calcium absorption in adolescents. The authors reported that frequent consumption of prebiotics in the diet by adolescents results in an increase in the ability of absorption of calcium in the human body and also aids in the process of mineralization. Apart from the better absorption of minerals, prebiotics also help in easy emptying of the stomach. Thus prebiotics help in eradicating the issues related to constipation which is a frequent complaint from individuals of all age group (Tateyama et al. 2005). Prebiotics show a favorable purgative effect in humans with the problem of constipation, thereby intensifying the frequency of stool and facilitating easy passage through the colon in a short time. These advantages of prebiotics are dependent on the age, and medical conditions of the individuals and it may vary from individual to individual, i.e., the effects may be lesser or it may be higher depending on the physical health and presence of the population of probiotic cells in the gut of a person consuming prebiotics. To enjoy the positive effect of prebiotics, they should be efficiently utilized by the probiotic cells present in the gut. The absence of these probiotics may hamper the positive effects of prebiotics in humans.

5.4.3 Regulation of Diabetes

Prebiotics are known to be used in the treatment of diabetes; the use of prebiotic inulin in managing diabetes goes back to 1925 (Yang et al. 2015). The meta-analysis of the animal model investigated the effect of fructooligosaccharide on the homeostasis of glucose. The intake of fructooligosaccharides reduced the fasting blood glucose level (Le Bourgot et al. 2018). In one of the trials, a consumption of 10 g fructooligosaccharide by a woman with diabetes for 60 days resulted in a reduced plasma glycemia level during fasting. Further, there was a significant enhancement in the activity of superoxide dismutase and antioxidant activity in the human models fed with fructooligosaccharide as compared with the control (Dehghan et al. 2014a; Gargari et al. 2013). The study suggested that the regular intake of prebiotics by the diabetic patient can positively regulate the blood sugar level, lower endotoxemia, and manage the inflammation. Similarly, in another human study prebiotic was given to an obese woman having diabetes (type II) for 56 days. The authors reported a significant lowering in the level of fasting glucose of 9.5% as compared with the control group (Dehghan et al. 2014b). The production of short-chain fatty acid is responsible for this prebiotic effect; the propionate formed as a prebiotic fermentation metabolite inhibits the process of glucose formation and stimulates the process of glucose breakdown, and it also reduces the formation of glucose from other sources by decreasing the fatty acids present in plasma. The other metabolite formed is butyrate which stimulates the process of formation of glucagon-like peptide (GLP-1) which manages blood glucose and also helps in managing obesity by increasing the feeling of hunger and thus reducing the appetite. Human studies are available for fructooligosaccharide and inulin; still, there is immense future scope to explore this area of research and investigate the effect of using different prebiotics in the diet on the males/females of different age groups suffering from diabetes.

5.4.4 *Maintains Plasma Serum Cholesterol*

The probiotic bacteria which grow and survive by extracting nutrition from the prebiotics are directly related to maintaining body weight and preventing obesity and related health issues (Dewulf et al. 2011; Włodarczyk and Śliżewska 2021). There are many studies on animal models demonstrating the beneficial effects of prebiotic components on lowering the plasma cholesterol level and reducing the risk of steatosis. Van der Beek et al. (2019) studied the effect of inulin consumption by individuals (20–50 years of age) on their metabolic health and oxidation of fat. The authors gave milkshakes containing 24 g inulin to the individuals and the sample was collected up to 7 h post-consumption; the investigation showed that there was an increase in the oxidation of fat in the early phase after consumption and a decrease in the serum glucose and plasma insulin levels. There was an increase in the concentration of acetate in blood plasma, which was the metabolite generated after the fermentation of inulin and there were no changes observed in the plasma triglyceride levels. Wong et al. (2010) studied the effect of prebiotics on plasma lipid profiles. The prebiotic component was added to soy food and this diet was given to a group of individuals which included 11 male and 12 female subjects having hyperlipidemia. The authors reported that there was a decrease in the low-density lipoprotein fraction and a significant increase in the high-density lipoprotein fraction. In some recent studies, the beneficial effect of the formation of short-chain fatty acid by colonic bacteria fermentation has shown to be beneficial in cholesterol management. Wu et al. (2020a, b) studied the effect of prebiotics from bamboo in lowering the cholesterol level. The author reported that the presence of dietary fibers in bamboo showed higher cholesterol-lowering capacities, i.e., insoluble and soluble dietary fibers from *Fargesia spathacea* species of bamboo showed 10.82 and 9.59 mg/g as adsorption capacities for cholesterol. Further authors reported that in vitro fermentation study showed that both soluble and insoluble dietary fibers have a positive effect on the population of *Bifidobacteria* and *Lactobacillus* species and increase the short-chain fatty concentration. Prebiotic consumption reduces hepatic de novo lipogenesis. Decrease in the synthesis of fat results in lowering the formation of very-low-density lipoprotein into the blood from the liver. The short-chain fatty acid formed as a metabolite can be correlated with a reduction in the cholesterol level. The acetate stimulated the process of cholesterogenesis and it is also known to promote the process of lipogenesis, whereas the other metabolite propionate restricts acetate to enter the hepatocyte and thus acts as an obstacle in the pathways for lipid and cholesterol synthesis and also propionate blocks the activity of hydroxymethylglutaryl coenzyme A, which is needed in the body to catalyze the formation of cholesterol. Blocking its activity restricts the formation and accumulation of cholesterol in the body and prevents various cholesterol-induced heart diseases. Hence, the formation of the fermentation end products like short-chain fatty acids has a beneficial effect on lowering cholesterol levels and maintaining blood lipid levels. So, the amount of acetate and propionate formed as a metabolite after fermentation can decide the potential to lower the lipid and cholesterol level in serum and also prevents obesity.

5.4.5 Prevention of Cancer

Cancer is recognized as a disease that is a major cause of death across the globe. The metabolites formed after the fermentation of prebiotics are of great importance in preventing colon cancer. The fermentation of prebiotics by the gut microflora produces short-chain fatty acids like butyrate, acetate, and propionate that reduced the pH in the gut and is being closely related to preventing the risk of cancer in humans (Mahdavi et al. 2021). The prebiotic intake should be high in diet to increase the fermentation and production of metabolites. Butyrate decreases cell proliferation by inhibiting the process of DNA synthesis (Demigne et al. 1999). The prebiotics restricts the formation of ammonia which modifies the morphology of cells in the intestine and also increases the synthesis of DNA, thus decreasing the survival of the mucosal cell. Hence, restricting the formation of ammonia is beneficial in preventing colon cancer (Hawrelak 2020). The fermentation of prebiotics also promotes the conversion of lignans into their active form and this active form of lignans and isoflavonoids together are known to have benefits in preventing cancer and also diminish the risk of breast and prostate cancer by inhibiting the effect of sex hormones (Kührer 2006; Pool-Zobel et al. 2000). In a human study, Ballongue et al. (1997) showed inhibition of the activity of numerous biologically active enzymes responsible for colon cancer by consumption of the prebiotic component lactulose. The authors reported that the lactulose decreased the activity of urease, 7-α-dehydroxylase, azoreductase, and β-glucuronidase by 25%, 40%, 45%, and 38%, respectively, thus reducing the risk of cancers in humans. There is future scope to investigate the potential of prebiotics in reducing the risk of other cancers through in vivo and in vitro studies, to make this claim more concrete. There are numerous prebiotics but still there is no data available in the literature for the specific effect of the individual prebiotics on the prevention of cancers in humans. Future research should focus on the incorporation of prebiotics in foods like biscuits, cookies, beverages, etc., and their effect on preventing the risk of cancer in humans. Food as a preventive measure against cancer is always better than prebiotics going in the form of tablets, drugs, syrups, etc.

5.4.6 Other Health Benefits

The consumption of prebiotics is closely correlated to the satiety value. A study by Cani et al. (2006) investigated the beneficial effects of prebiotics on satiety in humans. The subjects were given 8 g fructooligosaccharide for breakfast and dinner to 10 individuals in an age group of 20–40 years for 14 days. The authors reported that consumption of fructooligosaccharide enhanced satiety irrespective of age and the intake of food was also reduced during dinner. The energy intake was reduced by 5% in these individuals and also helped in maintaining body weight. One more study on humans by Parnell and Reimer (2009) demonstrated the positive effects of prebiotic consumption in overweight individuals. Overweight individuals were selected and they were given 21 g of prebiotic for 12 weeks and their effect was investigated. The authors reported that there was a reduction in the body weight of individuals by 1.03 kg consuming prebiotics whereas the body weight of the control group increased by 0.45 kg. Further, the intake of prebiotics resulted in improving blood glucose and insulin levels in these overweight individuals.

Prebiotics are also beneficial for skin health, though there is limited scientific evidence available in this area. Since the skin type of individuals from different geographic regions differs, there could be the possibility of differences in the effect of prebiotics on the skin of individuals from different geographic origins. Jung et al. (2017) investigated the effect of the consumption of a mixture of prebiotics on the skin health of healthy women. The subjects were given 4.5 g/day of a prebiotic mixture containing lactulose and galactooligosaccharides for 56 days. The authors reported that subjects taking a prebiotic mixture showed a reduction in the length and the depth of wrinkles as measured by skin evaluation parameters. The wrinkles became worse in the control group after 56 days. The similarly beneficial effect of intake of galactooligosaccharides on the health of the skin was also reported by Hong et al. (2017). Though there are few studies available in the literature the mechanism behind the improvement of skin health with the consumption of prebiotic components is not clear and is still a topic of research. Some research also suggested the effectiveness of prebiotics against anxiety. In a rat model, it was found that the single prebiotic and combination of prebiotics are beneficial in reducing stress in rats by releasing corticosterone (Burokas et al. 2017). Other beneficial effects include positive modification of neuroresponse related to stress (Schmidt et al. 2015) and prevention of hepatic encephalopathy (Dhiman et al. 2000), urine tract infection (Clarke 1993), atopic eczema (Moro et al. 2006), infection in the gastrointestinal tract (Drakoularakou et al. 2010), and irritable bowel syndrome (Silk et al. 2009). Prebiotics are a boon for human health as they help cure and prevent various infections in humans.

5.5 Commercially Available Prebiotic Foods

Consumers are becoming more conscious about their health and are attracted to foods with health benefits. Due to numerous health benefits of prebiotics, they are being incorporated into processed food products meant for human consumption. Prebiotic ingredients can be incorporated in food products like biscuits, cakes, confectionery, fruit juices, cookies, squash, dessert, ice creams, yogurt, breakfast cereals, etc. The global requirement for prebiotics is growing approximately by more than 500,000 tons/per annum (Douglas and Sanders 2008). As per the market report, the size of the prebiotic market was US8.95 billion dollars in 2020 and it is expected to increase by 7% by 2028. The global market size for prebiotics is continuously experiencing an increase due to the demand for these products in the market. Products like inulin powder from chicory root, Fiber gummies, MS Prebiotic, Prebiotin, and Ensure plus fiber are commercially available prebiotic products on the market. Prebiotin contains oligofructose and inulin which are known to positively increase the number of health-promoting bacteria in the human colon. The diet in Western countries lacks dietary fibers, which necessitates the fibers to be incorporated into the diet for keeping healthy; to overcome this, MS Prebiotic came up with a product containing 70% resistant starch which helps in increasing microflora in the gut and maintains gastrointestinal health. It is commercially available on the market in Canada and the United States. Other commercially available products include Ensure plus fiber (EPF) which contains fructooligosaccharide; it is manufactured by Abbott Nutrition, USA. EPF is being marketed as a beneficial product against constipation, neurological disorders, low appetite, malnutrition, etc. This product is not suitable for growing kids and individuals suffering from galactosemia. Inulin powder manufactured by Jetsu technologies, USA, contains inulin as a prebiotic which is claimed to cure constipation, aid in process of digestion, and increase the population of microflora in the gut. It is not suitable for pregnant women and patients suffering from diabetes and gallstone. Another product from Abbott Nutrition suitable for children (above 1 year of age), adults, and elderly consumers is Ensure Twocal which contains fructooligosaccharide as a prebiotic component. Other commercially available products which maintain blood lipid level and plasma glucose level, maintain body weight, reduce low-density lipoprotein, and strengthen the immune system are Fiber Gummies and Jarrow formula. Fiber gummies contain inulin and chicory root and are manufactured by BeLive, USA. Jarrow formula contains both xylooligosaccharides and galactooligosaccharides manufactured by Jarrow, USA. Nowadays the local manufacturer also is incorporating prebiotics in processed food to capture the market and to make higher profits. There would be more products available on the market in the near future across the globe.

5.6 Safety of Prebiotics

Prebiotics are generally considered as safe for human consumption as they are obtained from the edible sources like cereals, pulses, fruits, vegetables, etc. There is evidence available from animal and human studies showing the beneficial effect of the consumption of prebiotics with little or no adverse effects. In general dosage of 4 g of fructooligosaccharide, 10 g of galactooligosaccharides, and 2 g of xylooligosaccharides per day is required to raise the population of beneficial microflora of the gut (Manning and Gibson 2015). Consumption of prebiotics beyond a normal level may have some negative impacts like constipation, enhanced gas production, etc. (Tuohy et al. 2003). This can be overcome by initially starting with the low doses of around 2 g per day; then it can be gradually increased as needed, to enjoy its beneficial effects. It is observed that adults can tolerate an initial intake of around 2–2.5 g/day with little gastrointestinal issues; these issues may be limited to some individuals. For the treatment of anxiety dosage up to 7 g/day, for prevention of diarrhea 2.6 g/day but for the absorption of minerals, the dosage of 20 g/day is suggested. The intake also varies with the conditions to be managed by prebiotics. Usage of galactooligosaccharides in food products for human consumption is generally safe. Many clinical studies show that galactooligosaccharides are well tolerated by individuals of all age groups including infants and also in individuals who cannot tolerate other prebiotics (Hawrelak 2020). Other prebiotics like lactulose, glucans, fructans, etc. are also considered safe for human consumption as they themselves are part of the food which we consume. Lactulose intake of 3 g per day is found to be beneficial in increasing the bifidogenic effect in humans (Terada et al. 1992). More intense effects of lactulose were observed with a higher intake of more than 10 g/day (Hawrelak 2020). Overall looking at the data available in the literature, the intake of prebiotics is safe and beneficial to human beings. It could create some issues initially, but proper management of intake dosage could overcome this issue. Hence, food rich in prebiotics should be frequently included in the diet, which will help to keep the body healthy which will directly help to keep mental health sound.

5.7 Synbiotic Food

The term synbiotics suggest a synergism and this is exclusively used to refer food products wherein the prebiotic ingredient is used to significantly increase the population of probiotic cells in the human gut. The word synbiotic was coined in the mid-1990s by Gibson and Roberfroid to explain the synergism between prebiotic and probiotic (Kumari et al. 2020). The concept of synbiotic emerged to overcome and increase the rate of survival of probiotic cells in the gut. The prebiotic is known to stimulate the survival and growth of beneficial microflora in the human gut; the synergism results in the formation of metabolites namely short-chain fatty acid which reduces the pH of the gut and prevents various diseases or reduces the risk of various infections or diseases by boosting immunity. Due to the health benefits of synbiotic, there has been tremendous growth in the research area on synbiotic since the last 15 years. The most important factor to be considered while formulation of a synbiotic food is selection of appropriate prebiotic and probiotic that exert glaring benefits on human health when administered individually (de Vrese and Schrezenmeir 2016). There are numerous possibilities of combination of prebiotic and probiotic to be used for the formulation of synbiotic food to enhance microflora in the human gut but the combination should work well and thus this is a promising tool in the formulation of synbiotic food. Possibilities of different probiotics to ferment prebiotics could be different; for instance fructooligosaccharide, lactulose, and galactooligosaccharides are well fermented by *Bifidobacterium lactis* but this bacteria cannot ferment lactitol (Vernazza et al., 2006). Similarly, *Lactobacillus rhamnosus* cannot ferment fructooligosaccharide, lactulose, and lactitol, but *Lactobacillus acidophilus* can ferment fructooligosaccharide (Datta et al. 2007; Hutkins and Kaplan 2000; Kontula et al. 1999). Therefore, it is very important to properly select a prebiotic and probiotic combination for formulating synbiotic food to enjoy the probable health benefits. The prebiotic ingredient used in formulating synbiotic food should be effectively utilized by the probiotics used in formulation and should promote growth and survival of probiotics to confer health benefits to the host. It should be taken care that the prebiotics used in formulation of synbiotic food should not promote the growth of pathogenic bacteria which will be harmful for human beings. There is no or very few research available in the literature focusing on the best combination of prebiotic and probiotic to be used for formulation of a synbiotic food.

5.7.1 Functioning of Synbiotic Food

The synergistic approach of prebiotic and probiotic is observed in the synbiotic food; it is known that the beneficial microflora is active in the intestinal region and effectively utilized prebiotic components to confer health benefits to the host. Probiotics selectively ferment prebiotic components, and these prebiotics serve as a basic medium for the proliferation of probiotic microflora in the gut. There is evidence available in the literature showing an increase in the tolerance of probiotic microflora to conditions like temperature, pH, and other adverse condition in the gut of an host consuming prebiotics (Kumari et al. 2020; Singu et al. 2020a). The pathway through which probiotics become more tolerant to such harsh conditions is not explained in the literature and needs further research in this area. Possibly, the beneficial effect of synbiotic food to the host could be because of proliferation of probiotic microflora in the gut with the usage of prebiotic components which forms the metabolites, that lowers the pH of the gut and reduces the viability of pathogenic microorganism in the gut and helps in maintaining bio-population in the gut of the host organism (de Vrese and Schrezenmeir 2016). Synergistic effect of prebiotics and probiotics leads to a decrease in the concentration of unwanted metabolites which may be carcinogenic in nature and increase in the concentration of acetate, butyrate, propionate, ketones, methyl acetate, and carbon disulfides which positively benefits the host organism. The salubrious effect of synbiotic food includes antiobesity, anti-inflammatory, anticarcinogenic, antidiabetic, etc.

5.7.2 Synbiotics and Health

Synbiotics have been found directly related to human health and incorporation of synbiotic in the diet reduces the risk of many diseases. The most extensively used prebiotic components include inulin, fructooligosaccharide, and galactooligosaccharides and the most commonly used probiotics include *Bifidobacteria and Lactobacillus species* strain in the formulation of synbiotic foods. These mentioned prebiotic and probiotic components are also among the most researched components about which clinical evidence is also available in literature. One such study investigated the effect of synbiotic in treating nonalcoholic steatohepatitis. The authors used inulin as a prebiotic, and several probiotics which included *Lactobacillus rhamnosus, Lactobacillus acidophilus, Lactobacillus delbrueckii* and *Bifidobacteria* were used as a synbiotic. Similarly other studies reported the use of prebiotic fructooligosaccharide and probiotic strain as *Lactobacillus acidophilus, Lactobacillus bulgaricus* and *Bifidobacteria.* This combination was effective against fatty liver diseases in nonalcoholic individuals (Eslamparast et al. 2014). The prebiotic components are naturally present in many foods as discussed in Sect. 5.3, of this chapter. The well-known food known to have synbiotic properties is breast milk; it contains oligosaccharides having prebiotic properties and lactic acid bacteria which are probiotics. The consumption of breast milk by infants confers numerous health benefits and helps in building immunity in infants to fight infections (Heyman and Ménard 2002; Wu et al. 2020a, b). Lately, there are formulations of infant formulas for infants who are not breastfed due to many unavoidable reasons; these infant formulas also contain oligosaccharides and probiotic strain to keep infants healthy (Sazawal et al. 2010). Prebiotic components like resistant starches are also present in many foods; raw banana contains the highest proportion of resistant starch in it which has health benefits like antiobesity, lowers serum cholesterol, reduces the risk of kidney diseases, etc. Raw banana could be a potential source for extraction of resistant starches for their use as a prebiotic ingredient in other synbiotic formulations. Batista et al. (2017) formulated synbiotic milk by using raw banana flour. The banana flour containing resistant starch was added to fermented milk to prepare a fermented synbiotic drink. The prebiotic resistant starch helps the lactic acid bacteria (probiotic) to grow and the population of these bacteria increased to 6 log CFU/g. The synbiotic effect of drink enhanced the triglyceride profile by increasing the essential fatty acid content and also increased the palatability of the food. The studies reported in the literature shows the beneficial effect of synbiotic against diarrhea (Dinleyici et al. 2013), bowel syndrome (Young and Cash 2006), pancreatitis (Forsmark and Baillie 2007), and infection in the respiratory tract (Parracho et al. 2007); it also prevents cancer (Cazzola et al. 2010) and positively affects hyperlipidaemia (Ooi et al. 2010).

5.8 Conclusion

Prebiotics and synbiotics have proved their health benefits and have attracted the attention of the food processing sector. They are now being incorporated in numerous processed foods and are becoming part of our diet. Prebiotics are naturally found in various sources like cereals, pulses, fruits, vegetables, etc. and prebiotics like fructooligosaccharide, xylooligosaccharides, fructans, glucans, etc. are among the most explored for their health benefits like antiobesity, antidiabetic, anti-inflammatory, anticarcinogenic etc. and these effects are more intensified with the combination of prebiotic components with the probiotic strains like *Bifidobacteria* spp. and *Lactobacillus* spp.; the combined effect of prebiotic and probiotic is popularly known as synbiotic. The specific combination of prebiotic components and probiotic strains is incorporated in food to increase the therapeutic value of food giving additional health benefits. The studies reported in the literature had shown that probiotic cells can selectively utilize prebiotic components and produce metabolites which lower the pH and decrease the population of pathogenic bacteria in the gut and bestow many health benefits to humans. However, there is a need for further studies on exploring better combination of prebiotic component and probiotic strain to widen the application routes and possible health benefits of synbiotic in foods.

Acknowledgment Harsh B. Jadhav would like to thank the Department of Science and Technology (DST), Ministry of Science and Technology, Government of India, for providing DST INSPIRE Fellowship.

Conflict of Interest The authors declare no conflict of interest.

References

Abdi R, Joye IJ (2021) Prebiotic potential of cereal components. Foods 10(10):1–18. https://doi.org/10.3390/foods10102338

Abeysekara S, Chilibeck PD, Vatanparast H, Zello GA (2012) A pulse-based diet is effective for reducing total and LDL-cholesterol in older adults. Br J Nutr 108(Suppl 1):S103–S110. https://doi.org/10.1017/S0007114512000748

Abrams SA, Griffin IJ, Hawthorne KM, Liang L, Gunn SK, Darlington G, Ellis KJ (2005) A combination of prebiotic short- and long-chain inulin-type fructans enhances calcium absorption and bone mineralization in young adolescents. Am J Clin Nutr 82(2):471–476. https://doi.org/10.1093/ajcn.82.2.471

Al-Ghazzewi FH, Tester RF (2014) Impact of prebiotics and probiotics on skin health. Benef Microbes 5(2):99–107. https://doi.org/10.3920/BM2013.0040

Anwaar A, Hafiz Muhammad RA, Ahmad A (2020) Utilization of mango peel in development of instant drink. Asian J Agric Biol 8(3):260–267. https://doi.org/10.35495/ajab.2020.02.094

Arena MP, Russo P, Capozzi V, Rascón A, Felis GE, Spano G, Fiocco D (2016) Combinations of cereal β-glucans and probiotics can enhance the anti-inflammatory activity on host cells by a synergistic effect. J Funct Foods 23:12–23. https://doi.org/10.1016/j.jff.2016.02.015

Arihara K, Ohata M (2011) Use of probiotics and prebiotics in meat products. In: Processed meats: improving safety, nutrition and quality. Woodhead Publishing Limited, Sawston. https://doi.org/10.1533/9780857092946.2.403

Ashwini A, Ramya HN, Ramkumar C, Reddy KR, Kulkarni RV, Abinaya V, Naveen S, Raghu AV (2019) Reactive mechanism and the applications of bioactive prebiotics for human health: review. J Microbiol Methods 159:128–137. https://doi.org/10.1016/j.mimet.2019.02.019

Ballongue J, Schumann C, Quignon P (1997) Effects of lactulose and lactitol on colonic microflora and enzymatic activity. Scand J Gastroenterol Suppl 32(222):41–44. https://doi.org/10.1080/00365521.1997.11720716

Balthazar CF, Conte Júnior CA, Moraes J, Costa MP, Raices RSL, Franco RM, Cruz AG, Silva ACO (2016) Physicochemical evaluation of sheep milk yogurts containing different levels of inulin. J Dairy Sci 99(6):4160–4168. https://doi.org/10.3168/jds.2015-10072

Balthazar CF, Silva HLA, Vieira AH, Neto RPC, Cappato LP, Coimbra PT, Moraes J, Andrade MM, Calado VMA, Granato D, Freitas MQ, Tavares MIB, Raices RSL, Silva MC, Cruz AG (2017) Assessing the effects of different prebiotic dietary oligosaccharides in sheep milk ice cream. Food Res Int 91:38–46. https://doi.org/10.1016/j.foodres.2016.11.008

Batista ALD, Silva R, Cappato LP, Ferreira MVS, Nascimento KO, Schmiele M, Esmerino EA, Balthazar CF, Silva HLA, Moraes J, Pimentel TC, Freitas MQ, Raices RSL, Silva MC, Cruz AG (2017) Developing a synbiotic fermented milk using probiotic bacteria and organic green banana flour. J Funct Foods 38:242–250. https://doi.org/10.1016/j.jff.2017.09.037

Baxter BA, Oppel RC, Ryan EP (2019) Navy beans impact the stool metabolome and metabolic pathways for colon health in cancer survivors. Nutrients 11(1):28. https://doi.org/10.3390/nu11010028

Ben K, Bouaziz F, Zouari-ellouzi S, Chaari F, Ellouz-chaabouni S, Ellouz-ghorbel R, Nouri-ellouz O (2017) Improvement of texture and sensory properties of cakes by addition of potato peel powder with high level of dietary fiber and protein. Food Chem 217:668–677. https://doi.org/10.1016/j.foodchem.2016.08.081

Beynen AC, Baas JC, Hoekemeijer PE, Kappert HJ, Bakker MH, Koopman JP, Lemmens AG (2002) Faecal bacterial profile, nitrogen excretion and mineral absorption in healthy dogs fed supplemental oligofructose. J Anim Physiol Anim Nutr 86(9–10):298–305. https://doi.org/10.1046/j.1439-0396.2002.00386.x

Bhagwat A, Bhushette P, Annapure US (2020) Spray drying studies of probiotic Enterococcus strains encapsulated with whey protein and maltodextrin. Beni-Suef Univ J Basic Appl Sci 9(1):33. https://doi.org/10.1186/s43088-020-00061-z

Biruete A, Cross TWL, Allen JM, Kistler BM, de Loor H, Evenepoel P, Fahey GC, Bauer L, Swanson KS, Wilund KR (2021) Effect of dietary inulin supplementation on the gut microbiota composition and derived metabolites of individuals undergoing hemodialysis: a pilot study. J Renal Nutr 31(5):512–522. https://doi.org/10.1053/j.jrn.2020.10.003

Bis-Souza CV, Penna ALB, da Silva Barretto AC (2020a) Applicability of potentially probiotic Lactobacillus casei in low-fat Italian type salami with added fructooligosaccharides: in vitro screening and technological evaluation. Meat Sci 168:108186. https://doi.org/10.1016/j.meatsci.2020.108186

Bis-Souza CV, Pateiro M, Domínguez R, Penna ALB, Lorenzo JM, Silva Barretto AC (2020b) Impact of fructooligosaccharides and probiotic strains on the quality parameters of low-fat Spanish Salchichón. Meat Sci 159:107936. https://doi.org/10.1016/j.meatsci.2019.107936

Blaut M (2002) Relationship of prebiotics and food to intestinal microflora. Eur J Nutr 41(Suppl 1):11–16. https://doi.org/10.1007/s00394-002-1102-7

Blottiere HM, Buecher B, Galmiche J-P, Cherbut C (2003) Molecular analysis of the effect of short-chain fatty acids on intestinal cell proliferation. Proc Nutr Soc 62(1):101–106. https://doi.org/10.1079/pns2002215

Bojarczuk A, Sk S, Mousavi A, Marsza K (2022) Health benefits of resistant starch: a review of the literature. J Funct 93:1–11. https://doi.org/10.1016/j.jff.2022.105094

Brummer Y, Kaviani M, Tosh SM (2015) Structural and functional characteristics of dietary fibre in beans, lentils, peas and chickpeas. Food Res Int 67:117–125. https://doi.org/10.1016/j.foodres.2014.11.009

Bultosa G, Dawa D (2019) Functional foods: dietary fibers, prebiotics, probiotics, and synbiotics. In: Encyclopedia of food grains, 2nd edn. Elsevier, Amsterdam, pp 11–16. https://doi.org/10.1016/B978-0-12-394437-5.00245-X

Burokas A, Arboleya S, Moloney RD, Peterson VL, Murphy K, Clarke G, Stanton C, Dinan TG, Cryan JF (2017) Targeting the microbiota-gut-brain axis: prebiotics have anxiolytic and antidepressant-like effects and reverse the impact of chronic stress in mice. Biol Psychiatry 82(7):472–487. https://doi.org/10.1016/j.biopsych.2016.12.031

Cani PD, Joly E, Horsmans Y, Delzenne NM (2006) Oligofructose promotes satiety in healthy human: a pilot study. Eur J Clin Nutr 60(5):567–572. https://doi.org/10.1038/sj.ejcn.1602350

Casiraghi MC, Zanchi R, Canzi E, Pagani MA, Viaro T, Benini L, D'Egidio MG (2011) Prebiotic potential and gastrointestinal effects of immature wheat grain (IWG) biscuits. Antonie van Leeuwenhoek 99(4):795–805. https://doi.org/10.1007/s10482-011-9553-6

Cazzola M, Pham-Thi N, Kerihuel JC, Durand H, Bohbot S (2010) Efficacy of a synbiotic supplementation in the prevention of common winter diseases in children: a randomized, double-blind, placebo-controlled pilot study. Ther Adv Respir Dis 4(5):271–278. https://doi.org/10.1177/1753465810379010

Cezar M, Mano R, Neri-numa IA, Bueno J, Paulino BN, Pessoa MG, Pastore GM (2017) Oligosaccharide biotechnology: an approach of prebiotic revolution on the industry. Appl Microbiol Biotechnol 102(1):17–37

Chan S, Hawley CM, Pascoe EM, Cao C, Campbell SB, Campbell KL, Francis RS, Hale R, Isbel NM, Morrison M, Johnson DW (2022) Prebiotic supplementation in kidney transplant recipients for preventing infections and gastrointestinal upset: a randomized controlled feasibility study. J Ren Nutr. https://doi.org/10.1053/j.jrn.2022.02.006

Clarisa Zepeda-Ruiz G, Abraham Domínguez-Avila J, Fernando Ayala-Zavala J, Robles-Sánchez M, Salazar-López NJ, Lopez A, Aguilar GAG (2020) Supplementing corn chips with mango cv. 'Ataulfo' peel improves their sensory acceptability and phenolic profile, and decreases. J Food Process Preserv 44(12). https://doi.org/10.1111/jfpp.14954

Clarke P (1993) Oral cholera vaccine well tolerated. BMJ 307:1425–1426

Coelho SR, Lima ÍA, Martins ML, Benevenuto Júnior AA, de Torres Filho R, de Lemos Souza Ramos A, Ramos EM (2019) Application of Lactobacillus paracasei LPC02 and lactulose as a potential symbiotic system in the manufacture of dry-fermented sausage. LWT 102(2019): 254–259. https://doi.org/10.1016/j.lwt.2018.12.045

Colantonio AG, Werner SL, Brown M (2020) The effects of prebiotics and substances with prebiotic properties on metabolic and inflammatory biomarkers in individuals with type 2 diabetes mellitus: a systematic review. J Acad Nutr Diet 120(4):587–607.e2. https://doi.org/10.1016/j.jand.2018.12.013

Costa MP, Frasao BS, Silva ACO, Freitas MQ, Franco RM, Conte-Junior CA (2015) Cupuassu (Theobroma grandiflorum) pulp, probiotic, and prebiotic: influence on color, apparent viscosity, and texture of goat milk yogurts. J Dairy Sci 98(9):5995–6003. https://doi.org/10.3168/jds.2015-9738

Costa GT, Vasconcelos QDJS, Abreu GC, Albuquerque AO, Vilar JL, Aragão GF (2020) Systematic review of the ingestion of fructooligosaccharides on the absorption of minerals and trace elements versus control groups. Clin Nutr ESPEN 41:68–76. https://doi.org/10.1016/j.clnesp.2020.11.007

da Silva TF, Conti-Silva AC (2018) Potentiality of gluten-free chocolate cookies with added inulin/oligofructose: chemical, physical and sensory characterization. LWT 90:172–179. https://doi.org/10.1016/j.lwt.2017.12.031

Datta S, Timson D, Annapure U (2007) Antioxidant properties and global metabolite screening of the probiotic yeast Saccharomyces cerevisiae var. boulardii. J Organ Behav 28(3):303–325. https://doi.org/10.1002/jsfa.8147

De Morais EC (2016) Prebiotic addition in dairy products: processing and health benefits. In: Probiotics, prebiotics, and synbiotics, vol 10. Elsevier, Amsterdam. https://doi.org/10.1016/B978-0-12-802189-7.00003-4

de Vrese M, Schrezenmeir J (2016) Probiotics, prebiotics and synbiotics. In: Functional foods: sources, health effects and future perspectives. Nova Science, Hauppauge, pp 143–207. https://doi.org/10.1201/b15561-2

Dehghan P, Gargari BP, Jafar-Abadi MA, Aliasgharzadeh A (2014a) Inulin controls inflammation and metabolic endotoxemia in women with type 2 diabetes mellitus: a randomized-controlled clinical trial. Int J Food Sci Nutr 65(1):117–123. https://doi.org/10.3109/09637486.2013.836738

Dehghan P, Pourghassem Gargari B, Asghari Jafar-abadi M (2014b) Oligofructose-enriched inulin improves some inflammatory markers and metabolic endotoxemia in women with type 2 diabetes mellitus: a randomized controlled clinical trial. Nutrition 30(4):418–423. https://doi.org/10.1016/j.nut.2013.09.005

Delgado-Fernández P, Corzo N, Olano A, Hernández-Hernández O, Moreno FJ (2019) Effect of selected prebiotics on the growth of lactic acid bacteria and physicochemical properties of yoghurts. Int Dairy J 89:77–85. https://doi.org/10.1016/j.idairyj.2018.09.003

Delgado-Fernández P, Moreno FJ, Corzo N, Nöbel S (2020) Physical properties of synbiotic yogurts as affected by the acidification rate. Int Dairy J 105:104665. https://doi.org/10.1016/j.idairyj.2020.104665

Demigne C, Remesy C, Morand C (1999) Short chain fatty acids. In: Colonic microbiota, nutrition and health. Springer, Dordrecht, pp 55–69. https://doi.org/10.1007/978-94-017-1079-4_4

Demirci T, Öztürk Negiş Hİ, Oraç A, Konak Göktepe Ç, Sözeri Atik D, Aktaş K, Demirci S, Sert D, Akın N (2019) Immature wheat grain as a potential prebiotic ingredient in set-type yoghurts: impact on antioxidative, textural properties and survival of different probiotics. J Food Sci Technol 56(12):5474–5483. https://doi.org/10.1007/s13197-019-04019-7

Dewulf EM, Cani PD, Neyrinck AM, Possemiers S, Van Holle A, Muccioli GG, Deldicque L, Bindels LB, Pachikian BD, Sohet FM, Mignolet E, Francaux M, Larondelle Y, Delzenne NM (2011) Inulin-type fructans with prebiotic properties counteract GPR43 overexpression and PPARγ-related adipogenesis in the white adipose tissue of high-fat diet-fed mice. J Nutr Biochem 22(8):712–722. https://doi.org/10.1016/j.jnutbio.2010.05.009

Dhiman RK, Sawhney IMS, Chawla YK, Das G, Ram S, Dilawari JB (2000) Efficacy of lactulose in cirrhotic patients with subclinical hepatic encephalopathy. Dig Dis Sci 45(8):1549–1552. https://doi.org/10.1023/A:1005556826152

Ding T, Li Y (2021) Beneficial effect and mechanism of walnut oligopeptide on Lactobacillus plantarum Z7. Food Sci Nutr 9(2):672–681. https://doi.org/10.1002/fsn3.2029

Dinleyici EC, Dalgic N, Guven S, Ozen M, Kara A, Metin-Timur VAO, Sancar M, Kurugol Z, Tanir G, Ozturk D, Aydogdu S, Tutanc M, Eren M, Vandenplas Y (2013) The effect of a multispecies synbiotic mixture on the duration of diarrhea and length of hospital stay in children with acute diarrhea in Turkey: single blinded randomized study. Eur J Pediatr 172(4):459–464. https://doi.org/10.1007/s00431-012-1903-5

Domingo JL, Nadal M (2017) Carcinogenicity of consumption of red meat and processed meat: a review of scientific news since the IARC decision. Food Chem Toxicol 105:256–261. https://doi.org/10.1016/j.fct.2017.04.028

Douglas LC, Sanders ME (2008) Probiotics and prebiotics in dietetics practice. J Am Diet Assoc 108(3):510–521. https://doi.org/10.1016/j.jada.2007.12.009

Drakoularakou A, Tzortzis G, Rastall RA, Gibson GR (2010) A double-blind, placebo-controlled, randomized human study assessing the capacity of a novel galacto-oligosaccharide mixture in reducing travellers' diarrhoea. Eur J Clin Nutr 64(2):146–152. https://doi.org/10.1038/ejcn.2009.120

Drewnowski A, Fulgoni VL, Young MK, Pitman S (2008) Nutrient-rich foods: applying nutrient navigation systems to improve public health. J Food Sci 73(9):222–228. https://doi.org/10.1111/j.1750-3841.2008.00963.x

Duque ALRF, Demarqui FM, Santoni MM, Zanelli CF, Adorno MAT, Milenkovic D, Mesa V, Sivieri K (2021) Effect of probiotic, prebiotic, and synbiotic on the gut microbiota of autistic children using an in vitro gut microbiome model. Food Res Int 149:110657. https://doi.org/10.1016/j.foodres.2021.110657

El Zeny T, Essa RY, Bisar BA, Metwalli SM (2019) Effect of using chicory roots powder as a fat replacer on beef burger quality. Slov Vet Res 56(Suppl 22):509–514. https://doi.org/10.26873/SVR-788-2019

Eslamparast T, Poustchi H, Zamani F, Sharafkhah M, Malekzadeh R, Hekmatdoost A (2014) Synbiotic supplementation in nonalcoholic fatty liver disease: a randomized, double-blind, placebo-controlled pilot study. Am J Clin Nutr 99(3):535–542. https://doi.org/10.3945/ajcn.113.068890

Eslinger AJ, Eller LK, Reimer RA (2014) Yellow pea fiber improves glycemia and reduces Clostridium leptum in diet-induced obese rats. Nutr Res 34(8):714–722. https://doi.org/10.1016/j.nutres.2014.07.016

Fernando WMU, Hill JE, Zello GA, Tyler RT, Dahl WJ, Van Kessel AG (2010) Diets supplemented with chickpea or its main oligosaccharide component raffinose modify faecal microbial composition in healthy adults. Benef Microbes 1(2):197–207. https://doi.org/10.3920/BM2009.0027

Fincher GB, Stone BA (1974) A water-soluble arabinogalactan-peptide from wheat endosperm. Austr J Biol Sci 27(2):117–132. https://doi.org/10.1071/BI9740117

Finley JW, Burrell JB, Reeves PG (2007) Pinto bean consumption changes SCFA profiles in fecal fermentations, bacterial populations of the lower bowel, and lipid profiles in blood of humans. J Nutr 137(11):2391–2398. https://doi.org/10.1093/jn/137.11.2391

Fleita D, El-Sayed M, Rifaat D (2015) Evaluation of the antioxidant activity of enzymatically-hydrolyzed sulfated polysaccharides extracted from red algae; Pterocladia capillacea. LWT 63(2):1236–1244. https://doi.org/10.1016/j.lwt.2015.04.024

Forsmark CE, Baillie J (2007) AGA institute technical review on acute pancreatitis. Gastroenterology 132(5):2022–2044. https://doi.org/10.1053/j.gastro.2007.03.065

Garcia-Amezquita LE, Tejada-Ortigoza V, Serna-Saldivar SO, Welti-Chanes J, Garcia-Amezquita LE (2018) Dietary fiber concentrates from fruit and vegetable by-products: processing , modification , and application as functional ingredients association of official analytical chemists. Food Bioprocess Technol 11:1439–1463

Gargari BP, Dehghan P, Aliasgharzadeh A, Jafar-Abadi MA (2013) Effects of high performance inulin supplementation on glycemic control and antioxidant status in women with type 2 diabetes. Diabetes Metab J 37(2):140–148. https://doi.org/10.4093/dmj.2013.37.2.140

Gebruers K, Dornez E, Bedõ Z, Rakszegi M, Frás A, Boros D, Courtin CM, Delcour JA (2010) Environment and genotype effects on the content of dietary fiber and its components in wheat in the HEALTHGRAIN diversity screen. J Agric Food Chem 58(17):9353–9361. https://doi.org/10.1021/jf100447g

Gibson GR, Roberfroid MB (1995) Dietary modulation of the human colonic microbiota: introducing the concept of prebiotics. J Nutr 125(6):1401–1412. https://doi.org/10.1093/jn/125.6.1401

Gizem B, Kilicli M, Bursa K, Said O, Palabiyik I, Gulcu M, Yaman M, Gunes R, Konar N (2020) Effect of grape pomace usage in chocolate spread formulation on textural, rheological and digestibility properties. LWT 138:110451. https://doi.org/10.1016/j.lwt.2020.110451

Glisic M, Baltic M, Glisic M, Trbovic D, Jokanovic M, Parunovic N, Dimitrijevic M, Suvajdzic B, Boskovic M, Vasilev D (2019) Inulin-based emulsion-filled gel as a fat replacer in prebiotic- and PUFA-enriched dry fermented sausages. Int J Food Sci Technol 54(3):787–797. https://doi.org/10.1111/ijfs.13996

Glykys J, Duquette E, Rahmati N, Duquette K, Staley KJ (2019) Mannitol decreases neocortical epileptiform activity during early brain development via cotransport of chloride and water. Neurobiol Dis 125:163–175. https://doi.org/10.1016/j.nbd.2019.01.024

Gómez-Ordóñez E, Jiménez-Escrig A, Rupérez P (2010) Dietary fibre and physicochemical properties of several edible seaweeds from the northwestern Spanish coast. Food Res Int 43(9):2289–2294. https://doi.org/10.1016/j.foodres.2010.08.005

Graf D, Monk JM, Lepp D, Wu W, McGillis L, Roberton K, Brummer Y, Tosh SM, Power KA (2019) Cooked red lentils dose-dependently modulate the colonic microenvironment in healthy C57Bl/6 male mice. Nutrients 11(8):1–21. https://doi.org/10.3390/nu11081853

Gül H, Şen H (2017) Effects of pomegranate seed flour on dough rheology and bread quality. CyTA J Food 15(2):1–7. https://doi.org/10.1080/19476337.2017.1327461

Gullón P, Gullón B, Tavaria F, Vasconcelos M, Gomes AM (2015) In vitro fermentation of lupin seeds (Lupinus albus) and broad beans (Vicia faba): dynamic modulation of the intestinal microbiota and metabolomic output. Food Funct 6(10):3316–3322. https://doi.org/10.1039/c5fo00675a

Gupta S, Abu-Ghannam N (2011) Bioactive potential and possible health effects of edible brown seaweeds. Trends Food Sci Technol 22(6):315–326. https://doi.org/10.1016/j.tifs.2011.03.011

Ha V, Sievenpiper JL, de Souza RJ, Jayalath VH, Mirrahimi A, Agarwal A, Chiavaroli L, Mejia SB, Sacks FM, Di Buono M, Bernstein AM, Leiter LA, Kris-Etherton PM, Vuksan V, Bazinet RP, Josse RG, Beyene J, Kendall CWC, Jenkins DJA (2014) Effect of dietary pulse intake on established therapeutic review and meta-analysis of randomized controlled trials. Can Med Assoc J 186(8):252–262. https://doi.org/10.1503/cmaj.131727/-/DC1

Harris S, Powers S, Monteagudo-Mera A, Kosik O, Lovegrove A, Shewry P, Charalampopoulos D (2020) Determination of the prebiotic activity of wheat arabinogalactan peptide (AGP) using batch culture fermentation. Eur J Nutr 59(1):297–307. https://doi.org/10.1007/s00394-019-01908-7

Hathwar SC, Rai AK, Modi VK, Narayan B (2012) Characteristics and consumer acceptance of healthier meat and meat product formulations—a review. J Food Sci Technol 49(6):653–664. https://doi.org/10.1007/s13197-011-0476-z

Hawrelak JA (2020) Prebiotics, synbiotics, and colonic foods. In: Textbook of natural medicine, 5th edn. Elsevier, Amsterdam. https://doi.org/10.1016/b978-0-323-43044-9.00104-7

Helkar PB, Sahoo AK, Patil NJ (2016) Review: food industry by-products used as a functional food ingredients. Int J Waste Resour 6(3):1–6. https://doi.org/10.4172/2252-5211.1000248

Hesam F, Tarzi BG, Honarvar M, Jahadi M (2021) Pistachio (Pistacia vera) shell as a new candidate for enzymatic production of xylooligosaccharides. J Food Meas Charact 15(1):33–45. https://doi.org/10.1007/s11694-020-00594-y

Heyman M, Ménard S (2002) Cellular and molecular life sciences probiotic microorganisms: how they affect intestinal pathophysiology. Cell Mol Life Sci 59(1):1151–1165

Holscher HD, Guetterman HM, Swanson KS, An R, Matthan NR, Lichtenstein AH, Novotny JA, Baer DJ (2018) Walnut consumption alters the gastrointestinal microbiota, microbially derived secondary bile acids, and health markers in healthy adults: a randomized controlled trial. J Nutr 148(6):861–867. https://doi.org/10.1093/jn/nxy004

Hong YH, Chang UJ, Kim YS, Jung EY, Suh HJ (2017) Dietary galacto-oligosaccharides improve skin health: a randomized double blind clinical trial. Asia Pac J Clin Nutr 26(4):613–618. https://doi.org/10.6133/apjcn.052016.05

Hutkins R, Kaplan H (2000) Metabolism of fructo-oligosaccharides by lactic acid bacteria and bifidobacteria. Appl Environ Microbiol 66(6):2682–2684. https://doi.org/10.1128/AEM.66.6.2682-2684.2000.Updated

Illippangama AU, Jayasena DD, Jo C, Mudannayake DC (2022) Inulin as a functional ingredient and their applications in meat products. Carbohydr Polym 275:118706. https://doi.org/10.1016/j.carbpol.2021.118706

Inés C, Orcid V, Inlain DLI, Nacional U, Consejo UNL (2020) Influence of carrot fiber powder addition on rheological, microstructure and sensory characteristics of stirred-type yogurt. Int J Food Sci Technol 55(5):1916–1923. https://doi.org/10.1111/ijfs.14415

Jovanovic-Malinovska R, Kuzmanova S, Winkelhausen E (2014) Oligosaccharide profile in fruits and vegetables as sources of prebiotics and functional foods. Int J Food Prop 17(5):949–965. https://doi.org/10.1080/10942912.2012.680221

Jovanovic-Malinovska R, Kuzmanova S, Winkelhausen E (2015) Application of ultrasound for enhanced extraction of prebiotic oligosaccharides from selected fruits and vegetables. Ultrason Sonochem 22:446–453. https://doi.org/10.1016/j.ultsonch.2014.07.016

Jung EY, Kwon JI, Hong YH, Suh HJ (2017) Evaluation of anti-wrinkle effects of duoligo, composed of lactulose and galactooligosaccharides. Prev Nutr Food Sci 22(4):381–384. https://doi.org/10.3746/pnf.2017.22.4.381

Kadyan S, Sharma A, Arjmandi B, Singh P, Nagpal R (2022) Prebiotic potential of dietary beans and pulses and their resistant starch for aging-associated gut and metabolic health. Nutrients 14(1726):1–26

Kehinde BA, Panghal A, Garg MK, Sharma P, Chhikara N (2020) Vegetable milk as probiotic and prebiotic foods. In: Advances in food and nutrition research, vol 94, 1st edn. Elsevier, Amsterdam. https://doi.org/10.1016/bs.afnr.2020.06.003

Keulers L, Dehghani A, Knippels L, Garssen J, Papadopoulos N, Folkerts G, Braber S, van Bergenhenegouwen J (2022) Probiotics, prebiotics, and synbiotics to prevent or combat air pollution consequences: the gut-lung axis. Environ Pollut 302:119066. https://doi.org/10.1016/j.envpol.2022.119066

Khalil NA, Eltahan NR, Elaktash HM, Aly S, Sarbini SR (2021) Prospective evaluation of probiotic and prebiotic supplementation on diabetic health associated with gut microbiota. Food Biosci 42:101149. https://doi.org/10.1016/j.fbio.2021.101149

Koca I, Tekguler B, Arif V, Koca AF (2017) The use of grape, pomegranate and rosehip seed flours in Turkish noodle (erişte) production. J Food Process Preserv 42(1):e13343. https://doi.org/10.1111/jfpp.13343

Kontula P, Suihko ML, Von Wright A, Mattila-Sandholm T (1999) The effect of lactose derivatives on intestinal lactic acid bacteria. J Dairy Sci 82(2):249–256. https://doi.org/10.3168/jds.S0022-0302(99)75230-6

Kührer I (2006) Nutrition in breast cancer prevention. Breast Care 1(5):306–309. https://doi.org/10.1159/000096347

Kumar V, Sinha AK, Makkar HPS, de Boeck G, Becker K (2012) Dietary roles of non-starch polysaccharides in human nutrition: a review. Crit Rev Food Sci Nutr 52(10):899–935. https://doi.org/10.1080/10408398.2010.512671

Kumari R, Singh A, Yadav AN, Mishra S, Sachan A, Sachan SG (2020) Probiotics, prebiotics, and synbiotics: current status and future uses for human health. In: New and future developments in microbial biotechnology and bioengineering. Elsevier, Amsterdam. https://doi.org/10.1016/b978-0-12-820528-0.00012-0

Lam K, Cheung PC (2013) Non-digestible long chain beta-glucans as novel prebiotics. Bioact Carbohyd Diet Fibre 2(1):45–64. https://doi.org/10.1016/j.bcdf.2013.09.001

Lange KW, Hauser J, Nakamura Y, Kanaya S (2015) Dietary seaweeds and obesity. Food Sci Hum Wellness 4(3):87–96. https://doi.org/10.1016/j.fshw.2015.08.001

Le Bourgot C, Apper E, Blat S, Respondek F (2018) Fructo-oligosaccharides and glucose homeostasis: a systematic review and meta-analysis in animal models. Nutr Metab 15(1):1–15. https://doi.org/10.1186/s12986-018-0245-3

Li T, Xu L, Yan Q, Liu J, Jiang Z (2022) Sucrose-free hawthorn leathers formulated with fructooligosaccharides and xylooligosaccharides ameliorate high-fat diet induced infl ammation, glucose and lipid metabolism in liver of mice. Food Sci Hum Wellness 11(4): 1064–1075. https://doi.org/10.1016/j.fshw.2022.03.033

Lippi G, Mattiuzzi C, Cervellin G (2016) Meat consumption and cancer risk: a critical review of published meta-analyses. Crit Rev Oncol Hematol 97:1–14. https://doi.org/10.1016/j.critrevonc.2015.11.008

Liu Z, Lin X, Huang G, Zhang W, Rao P, Ni L (2014) Prebiotic effects of almonds and almond skins on intestinal microbiota in healthy adult humans. Anaerobe 26:1–6. https://doi.org/10.1016/j.anaerobe.2013.11.007

Long J, Yang J, Henning SM, Woo SL, Hsu M, Chan B, Heber D, Li Z (2019) Xylooligosaccharide supplementation decreases visceral fat accumulation and modulates cecum microbiome in mice. J Funct Foods 52:138–146. https://doi.org/10.1016/j.jff.2018.10.035

Lutsiv T, Weir TL, McGinley JN, Neil ES, Wei Y, Thompson HJ (2021) Compositional changes of the high-fat diet-induced gut microbiota upon consumption of common pulses. Nutrients 13(11):3992. https://doi.org/10.3390/nu13113992

Mahdavi M, Laforest-Lapointe I, Massé E (2021) Preventing colorectal cancer through prebiotics. Microorganisms 9(6):1–16. https://doi.org/10.3390/microorganisms9061325

Mandalari G, Faulks RM, Bisignano C, Waldron KW, Narbad A, Wickham MSJ (2010) In vitro evaluation of the prebiotic properties of almond skins (Amygdalus communis L.). FEMS Microbiol Lett 304(2):116–122. https://doi.org/10.1111/j.1574-6968.2010.01898.x

Manning T, Gibson G (2015) Prebiotics. In: Encyclopedia of food and health, vol 18(2). Elsevier, Amsterdam, pp 464–471. https://doi.org/10.1016/B978-0-12-384947-2.00560-2
McGinley JN, Fitzgerald VK, Neil ES, Omerigic HM, Heuberger AL, Weir TL, McGee R, Vandemark G, Thompson HJ (2020) Pulse crop effects on gut microbial populations, intestinal function, and adiposity in a mouse model of diet-induced obesity. Nutrients 12(3):1–17. https://doi.org/10.3390/nu12030593
Menshova RV, Ermakova SP, Anastyuk SD, Isakov VV, Dubrovskaya YV, Kusaykin MI, Um BH, Zvyagintseva TN (2014) Structure, enzymatic transformation and anticancer activity of branched high molecular weight laminaran from brown alga Eisenia bicyclis. Carbohydr Polym 99:101–109. https://doi.org/10.1016/j.carbpol.2013.08.037
Mineo H, Amano M, Minaminida K, Chiji H, Shigematsu N, Tomita F, Hara H (2006) Two-week feeding of difructose anhydride III enhances calcium absorptive activity with epithelial cell proliferation in isolated rat cecal mucosa. Nutrition 22(3):312–320. https://doi.org/10.1016/j.nut.2005.06.015
Mollakhalili-Meybodi N, Arab M, Nematollahi A, Mousavi Khaneghah A (2021) Prebiotic wheat bread: technological, sensorial and nutritional perspectives and challenges. LWT 149:111823. https://doi.org/10.1016/j.lwt.2021.111823
Montalvo-González E, Aguilar-Hernández G, Selene A, Ruiz-López II, Pérez-Silva A, Vivar-Vera MDLÁ (2018) Production, chemical, physical and technological properties of antioxidant dietary fiber from pineapple pomace and effect as ingredient in sausages. CyTA J Food 16(1): 831–839. https://doi.org/10.1080/19476337.2018.1465125
Moro G, Arslanoglu S, Stahl B, Jelinek J, Wahn U, Boehm G (2006) A mixture of prebiotic oligosaccharides reduces the incidence of atopic dermatitis during the first six months of age. Arch Dis Childhood 91(10):814–819. https://doi.org/10.1136/adc.2006.098251
Nissen L, di Carlo E, Gianotti A (2020) Prebiotic potential of hemp blended drinks fermented by probiotics. Food Res Int 131:109029. https://doi.org/10.1016/j.foodres.2020.109029
Nouri M, Nasehi B, Samavati V, Mehdizadeh SA (2017) Optimizing the effects of Persian gum and carrot pomace powder for development of low-fat donut with high fibre content. Bioact Carbohydr Diet Fibre 9:39–45. https://doi.org/10.1016/j.bcdf.2017.01.001
Ohta A, Yamada S, Sasa R (2000) Fructooligosaccharide consumption enhances femoral bone volume and mineral concentrations in rats. J Nutr 130(7):1792–1795
Ooi LG, Bhat R, Rosma A, Yuen KH, Liong MT (2010) A synbiotic containing Lactobacillus acidophilus CHO-220 and inulin improves irregularity of red blood cells. J Dairy Sci 93(10): 4535–4544. https://doi.org/10.3168/jds.2010-3330
Özer D, Akin S, Özer B (2005) Effect of inulin and lactulose on survival of lactobacillus acidophilus LA-5 and bifidobacterium bifidum BB-02 in acidophilus-bifidus yoghurt. Food Sci Technol Int 11(1):19–24. https://doi.org/10.1177/1082013205051275
Padhi EMT, Ramdath DD (2017) A review of the relationship between pulse consumption and reduction of cardiovascular disease risk factors. J Funct Foods 38:635–643. https://doi.org/10.1016/j.jff.2017.03.043
Paesani C, Degano AL, Salvucci E, Zalosnik MI, Fabi JP, Sciarini LS, Perez GT (2020) Soluble arabinoxylans extracted from soft and hard wheat show a differential prebiotic effect in vitro and in vivo. J Cereal Sci 93:102956. https://doi.org/10.1016/j.jcs.2020.102956
Paiva L, Lima E, Patarra RF, Neto AI, Baptista J (2014) Edible Azorean macroalgae as source of rich nutrients with impact on human health. Food Chem 164:128–135. https://doi.org/10.1016/j.foodchem.2014.04.119
Parnell JA, Reimer RA (2009) Weight loss during oligofructose supplementation is associated with decreased ghrelin and increased peptide YY in overweight and obese adults. Am J Clin Nutr 89(6):1751–1759. https://doi.org/10.3945/ajcn.2009.27465
Parracho H, McCartney AL, Gibson GR (2007) Probiotics and prebiotics in infant nutrition. Proc Nutr Soc 66(3):405–411. https://doi.org/10.1017/S0029665107005678
Pasqualetti V, Altomare A, Guarino MPL, Locato V, Cocca S, Cimini S, Palma R, Alloni R, De Gara L, Cicala M (2014) Antioxidant activity of inulin and its role in the prevention of human

colonic muscle cell impairment induced by lipopolysaccharide mucosal exposure. PLoS One 9(5):e98031. https://doi.org/10.1371/journal.pone.0098031
Pena AS (2007) Intestinal flora, probiotics, prebiotics, synbiotics and novel foods. Revista Espanola De Enfermedades Digestivas 99(11):653–658
Pérez-burillo S, Mehta T, Pastoriza S, Kramer DL, Paliy O, Rufián-henares JÁ (2019) Potential probiotic salami with dietary fiber modulates antioxidant capacity, short chain fatty acid production and gut microbiota community structure. LWT 105(5):355–362
Peshev D, Van den Ende W (2014) Fructans: prebiotics and immunomodulators. J Funct Foods 8(1):348–357. https://doi.org/10.1016/j.jff.2014.04.005
Phebean IO, Akinyele O, Toyin A, Folasade O (2017) Development and quality evaluation of carrot powder and cowpea flour enriched biscuits. Int J Food Sci Biotechnol 3:67–72. https://doi.org/10.11648/j.ijfsb.20170203.15
Pokusaeva K, Fitzgerald GF, Van Sinderen D (2011) Carbohydrate metabolism in bifidobacteria. Genes Nutr 6(3):285–306. https://doi.org/10.1007/s12263-010-0206-6
Pool-Zobel BL, Adlercreutz H, Glei M, Liegibel UM, Sittlingon J, Rowland I, Wähälä K, Rechkemmer G (2000) Isoflavonoids and lignans have different potentials to modulate oxidative genetic damage in human colon cells. Carcinogenesis 21(6):1247–1252. https://doi.org/10.1093/carcin/21.6.1247
Pop C, Suharoschi R (2021) Dietary fiber and prebiotic compounds in fruits and vegetables food waste. Sustainability 13:1–18
Provost V, Lamantia V, Bissonnette S, Cyr Y, Faraj M (2019) Increased fiber intake predicts the decrease in 2nd phase glucose-induced hyperinsulinemia following a hypocaloric diet in obese subjects. Nutr Metab Cardiovasc Dis 29(5):504–512. https://doi.org/10.1016/j.numecd.2019.01.014
Qi H, Huang L, Liu X, Liu D, Zhang Q, Liu S (2012) Antihyperlipidemic activity of high sulfate content derivative of polysaccharide extracted from Ulva pertusa (Chlorophyta). Carbohydr Polym 87(2):1637–1640. https://doi.org/10.1016/j.carbpol.2011.09.073
Raschka L, Daniel H (2005) Mechanisms underlying the effects of inulin-type fructans on calcium absorption in the large intestine of rats. Bone 37(5):728–735. https://doi.org/10.1016/j.bone.2005.05.015
Ribeiro TB (2020) Are olive pomace powders a safe source of bioactives and nutrients? J Sci Food Agric 101(5):1963–1978. https://doi.org/10.1002/jsfa.10812
Roberfroid M, Gibson GR, Hoyles L, McCartney AL, Rastall R, Rowland I, Wolvers D, Watzl B, Szajewska H, Stahl B, Guarner F, Respondek F, Whelan K, Coxam V, Davicco MJ, Léotoing L, Wittrant Y, Delzenne NM, Cani PD et al (2019) Prebiotic effects: metabolic and health benefits. Br J Nutr 104(Suppl 2):S1–S63. https://doi.org/10.1017/S0007114510003363
Rosa MC, Carmo MRS, Balthazar CF, Guimarães JT, Esmerino EA, Freitas MQ, Silva MC, Pimentel TC, Cruz AG (2021) Dairy products with prebiotics: an overview of the health benefits, technological and sensory properties. Int Dairy J 117:105009. https://doi.org/10.1016/j.idairyj.2021.105009
Samanta AK, Kolte AP, Senani S, Sridhar M, Jayapal N (2011) Prebiotics in ancient Indian diets. Curr Sci 101(1):43–46
Sazawal S, Dhingra U, Hiremath G, Sarkar A, Dhingra P, Dutta A, Verma P, Menon VP, Black RE (2010) Prebiotic and probiotic fortified milk in prevention of morbidities among children: community-based, randomized, double-blind, controlled trial. PLoS One 5(8):e12164. https://doi.org/10.1371/journal.pone.0012164
Schmidt K, Cowen PJ, Harmer CJ, Tzortzis G, Errington S, Burnet PWJ (2015) Prebiotic intake reduces the waking cortisol response and alters emotional bias in healthy volunteers. Psychopharmacology 232(10):1793–1801. https://doi.org/10.1007/s00213-014-3810-0
Scholz-Ahrens KE, Schaafsma G, van den Heuvel EGHM, Schrezenmeir J (2008) Effects of prebiotics on mineral absorption. Am J Clin Nutr 73(2):459–464. https://doi.org/10.1201/9780849381829.ch5

Scott KP, Grimaldi R, Cunningham M, Sarbini SR, Wijeyesekera A, Tang MLK, Lee JCY, Yau YF, Ansell J, Theis S, Yang K, Menon R, Arfsten J, Manurung S, Gourineni V, Gibson GR (2020) Developments in understanding and applying prebiotics in research and practice—an ISAPP conference paper. J Appl Microbiol 128(4):934–949. https://doi.org/10.1111/jam.14424

Shafi A, Naeem Raja H, Farooq U, Akram K, Hayat Z, Naz A, Nadeem HR (2019) Antimicrobial and antidiabetic potential of synbiotic fermented milk: a functional dairy product. Int J Dairy Technol 72(1):15–22. https://doi.org/10.1111/1471-0307.12555

Sheflin AM, Borresen EC, Kirkwood JS, Boot CM, Whitney AK, Lu S, Brown RJ, Broeckling CD, Ryan EP, Weir TL (2017) Dietary supplementation with rice bran or navy bean alters gut bacterial metabolism in colorectal cancer survivors. Mol Nutr Food Res 61(1):1–11. https://doi.org/10.1002/mnfr.201500905

Sievenpiper JL, Kendall CWC, Esfahani A, Wong JMW, Carleton AJ, Jiang HY, Bazinet RP, Vidgen E, Jenkins DJA (2009) Effect of non-oil-seed pulses on glycaemic control: a systematic review and meta-analysis of randomised controlled experimental trials in people with and without diabetes. Diabetologia 52(8):1479–1495. https://doi.org/10.1007/s00125-009-1395-7

Silk DBA, Davis A, Vulevic J, Tzortzis G, Gibson GR (2009) Clinical trial: the effects of a trans-galactooligosaccharide prebiotic on faecal microbiota and symptoms in irritable bowel syndrome. Aliment Pharmacol Ther 29(5):508–518. https://doi.org/10.1111/j.1365-2036.2008.03911.x

Sima P, Vannucci L, Vetvicka V (2018) β-glucans and cholesterol (review). Int J Mol Med 41: 1799–1808. https://doi.org/10.3892/ijmm.2018.3411

Singh AK, Cabral C, Kumar R, Ganguly R, Pandey AK (2020) Beneficial effects of dietary polyphenols on gut microbiota and strategies to improve delivery efficiency. Nutrients 11(2216):1–21

Singu BD, Bhushette PR, Annapure US (2020a) Survivability assessment of saccharomyces boulardii in a symbiotic system using nutraceuticals and modified atmosphere packaging. Food Bioprocess Technol 13(4):693–704. https://doi.org/10.1007/s11947-020-02430-z

Singu BD, Bhushette PR, Annapure US (2020b) Thermo-tolerant Saccharomyces cerevisiae var. boulardii coated cornflakes as a potential probiotic vehicle. Food Biosci 36:100668. https://doi.org/10.1016/j.fbio.2020.100668

Sisconeto Bisinotto M, da Silva DC, de Carvalho Fino L, Moreira Simabuco F, Neves Bezerra RM, Costa Antunes AE, Bertoldo Pacheco MT (2021) Bioaccessibility of cashew nut kernel flour compounds released after simulated in vitro human gastrointestinal digestion. Food Res Int 139: 109906. https://doi.org/10.1016/j.foodres.2020.109906

Skinner RC, Gigliotti JC, Ku K, Tou JC (2018) A comprehensive analysis of the composition, health benefits, and safety of apple pomace. Nutr Rev 8:1–17. https://doi.org/10.1093/nutrit/nuy033

Souza FP, Balthazar CF, Guimarães JT, Pimentel TC, Esmerino EA, Freitas MQ, Raices RSL, Silva MC, Cruz AG (2019) The addition of xyloligoosaccharide in strawberry-flavored whey beverage. LWT 109:118–122. https://doi.org/10.1016/j.lwt.2019.03.093

Speranza B, Campaniello D, Monacis N, Bevilacqua A, Sinigaglia M, Corbo MR (2018) Functional cream cheese supplemented with Bifidobacterium animalis subsp. lactis DSM 10140 and Lactobacillus reuteri DSM 20016 and prebiotics. Food Microbiol 72:16–22. https://doi.org/10.1016/j.fm.2017.11.001

Staneva G, Petkova D, Engeniring B (2014) Beta-glucans alter cholesterol level and susceptibility to oxidation in rat hepatocytes. Comptes Rendus de l'Académie Bulgare Des Sciences 67(10): 1383–1386

Sun Q, Zhu L, Li Y, Cui Y, Jiang S, Tao N, Chen H, Zhao Z, Xu J, Dong C (2020) A novel inulin-type fructan from Asparagus cochinchinensis and its beneficial impact on human intestinal microbiota. Carbohydr Polym 247:116761. https://doi.org/10.1016/j.carbpol.2020.116761

Surek E, Buyukkileci AO (2017) Production of xylooligosaccharides by autohydrolysis of hazelnut (Corylus avellana L.) shell. Carbohydr Polym 174:565–571. https://doi.org/10.1016/j.carbpol.2017.06.109

Talukder S (2015) Effect of dietary fiber on properties and acceptance of meat products: a review. Crit Rev Food Sci Nutr 55(7):1005–1011. https://doi.org/10.1080/10408398.2012.682230

Tateyama I, Hashii K, Johno I, Iino T, Hirai K, Suwa Y, Kiso Y (2005) Effect of xylooligosaccharide intake on severe constipation in pregnant women. J Nutr Sci Vitaminol 51(6):445–448. https://doi.org/10.3177/jnsv.51.445

Terada A, Hara H, Kataoka M, Mitsuoka T (1992) Effect of lactulose on the composition and metabolic activity of the human faecal flora. Microb Ecol Health Dis 5(1):43–50. https://doi.org/10.3109/08910609209141303

Thøgersen R, Castro-Mejía JL, Sundekilde UK, Hansen LH, Hansen AK, Nielsen DS, Bertram HC (2019) Ingestion of an inulin-enriched pork sausage product positively modulates the gut microbiome and metabolome of healthy rats. Mol Nutr Food Res 62(19):1–32. https://doi.org/10.1002/mnfr.201800608

Thompson HJ, Brick MA (2016) Perspective: closing the dietary fiber gap: an ancient solution for a 21st century problem. Adv Nutr 7(4):623–626. https://doi.org/10.3945/an.115.009696

Trinidad TP, Wolever TMS, Thompson LU (1996) Effect of acetate and propionate on calcium absorption from the rectum and distal colon of humans. Am J Clin Nutr 63(4):574–578. https://doi.org/10.1093/ajcn/63.4.574

Tunçil YE (2020) Dietary fibre profiles of Turkish Tombul hazelnut (Corylus avellana L.) and hazelnut skin. Food Chem 316:126338. https://doi.org/10.1016/j.foodchem.2020.126338

Tungland T (2018) Nondigestible fructans as prebiotic. In: Human microbiota in health and diseases (Issue 2007), pp 349–379. https://doi.org/10.1016/B978-0-12-814649-1/00008-9

Tuohy KM, Probert HM, Smejka CW, Gibson GR (2003) Using probiotics and prebiotics to improve gut health. Drug Discov Today 8(15):692–700

Tymczyszyn EE, Santos MI, Costa MC, Illanes A, Gómez-Zavaglia A (2014) Chapter 5: History, synthesis, properties, applications and regulatory issues of prebiotic oligosaccharides. In: Carbohydrated application in medicine (Vol 37661, Issue 2, pp 127–154). https://www.trnres.com/ebook/uploads/gilcontent/T_1401355846gil5.pdf

Udayarajan CT, Mohan K, Nisha P (2022) Tree nuts: treasure mine for prebiotic and probiotic dairy free vegan products. Trends Food Sci Technol 124:208–218. https://doi.org/10.1016/j.tifs.2022.04.022

Ursachi CŞ, Perţa-Crişan S, Munteanu FD (2020) Strategies to improve meat products' quality. Foods 9(12):1883. https://doi.org/10.3390/foods9121883

Valencia MS, Salgado SM, Andrade SAC, Padilha VM, Livera AVS, Stamford TLM (2016) Development of creamy milk chocolate dessert added with fructo-oligosaccharide and Lactobacillus paracasei subsp. paracasei LBC 81. LWT 69:104–109. https://doi.org/10.1016/j.lwt.2016.01.039

Vallath A, Shanmugam A, Rawson A (2022) Prospects of future pulse milk variants from other healthier pulses—as an alternative to soy milk. Trends Food Sci Technol 124:51–62. https://doi.org/10.1016/j.tifs.2022.03.028

Van der Beek CM, Canfora EE, Kip AM, Gorissen SHM, Olde Damink SWM, van Eijk HM, Holst JJ, Blaak EE, Dejong CHC, Lenaerts K (2019) The prebiotic inulin improves substrate metabolism and promotes short-chain fatty acid production in overweight to obese men. Metab Clin Exp 87:25–35. https://doi.org/10.1016/j.metabol.2018.06.009

Vernazza CL, Gibson GR, Rastall RA (2006) Carbohydrate preference, acid tolerance and bile tolerance in five strains of Bifidobacterium. J Appl Microbiol 100(4):846–853. https://doi.org/10.1111/j.1365-2672.2006.02832.x

Verspreet J, Dornez E, van den Ende W, Delcour JA, Courtin CM (2015) Cereal grain fructans: structure, variability and potential health effects. Trends Food Sci Technol 43(1):32–42. https://doi.org/10.1016/j.tifs.2015.01.006

Walton G, Swann J, Gibson G (2013) Prebiotics. In: The prokaryotes: human microbiology. Springer, Berlin, pp 1–554. https://doi.org/10.1007/978-3-642-30144-5

Wang S, Zhu H, Lu C, Kang Z, Luo Y, Feng L, Lu X (2012) Fermented milk supplemented with probiotics and prebiotics can effectively alter the intestinal microbiota and immunity of host animals. J Dairy Sci 95(9):4813–4822. https://doi.org/10.3168/jds.2012-5426

Wang X, Kristo E, Lapointe G (2020) Adding apple pomace as a functional ingredient in stirred-type yogurt and yogurt drinks. Food Hydrocolloids 100:105453. https://doi.org/10.1016/j.foodhyd.2019.105453

Wells ML, Potin P, Craigie JS, Raven JA, Merchant SS, Helliwell KE, Smith AG, Camire ME, Brawley SH (2017) Algae as nutritional and functional food sources: revisiting our understanding. J Appl Phycol 29(2):949–982. https://doi.org/10.1007/s10811-016-0974-5

Wen J-J, Li M-Z, Hu J-L, Tan H-Z, Nie S-P (2022) Resistant starches and gut microbiota. Food Chem 387:132895. https://doi.org/10.1016/j.foodchem.2022.132895

Wijesinghe WAJP, Jeon YJ (2012) Biological activities and potential industrial applications of fucose rich sulfated polysaccharides and fucoidans isolated from brown seaweeds: a review. Carbohydr Polym 88(1):13–20. https://doi.org/10.1016/j.carbpol.2011.12.029

Włodarczyk M, Śliżewska K (2021) Obesity as the 21st century's major disease: the role of probiotics and prebiotics in prevention and treatment. Food Biosci 42:101115. https://doi.org/10.1016/j.fbio.2021.101115

Wong JMW, Kendall CWC, De Souza R, Emam A, Marchie A, Vidgen E, Holmes C, Jenkins DJA (2010) The effect on the blood lipid profile of soy foods combined with a prebiotic: a randomized controlled trial. Metab Clin Exp 59(9):1331–1340. https://doi.org/10.1016/j.metabol.2009.12.017

Wong WY, Chan BD, Leung TW, Chen MX, Tai WCS (2022) Beneficial and anti-inflammatory effects of formulated prebiotics, probiotics, and synbiotics in normal and acute colitis mice. J Funct Foods 88:104871. https://doi.org/10.1016/j.jff.2021.104871

Wu RY, Johnson-Henry KC, Sherman PM (2020a) Prebiotics and human milk oligosaccharides. In: Encyclopedia of gastroenterology, 2nd edn. Elsevier, Amsterdam. https://doi.org/10.1016/b978-0-12-801238-3.66049-x

Wu W, Hu J, Gao H, Chen H, Fang X, Mu H, Han Y, Liu R (2020b) The potential cholesterol-lowering and prebiotic effects of bamboo shoot dietary fibers and their structural characteristics. Food Chem 332:127372. https://doi.org/10.1016/j.foodchem.2020.127372

Xavier-Santos D, Padilha M, Fabiano GA, Vinderola G, Cruz AG, Sivieri K, Costa Antunes AE (2022a) Corrigendum to 'Evidences and perspectives of the use of probiotics, prebiotics, synbiotics, and postbiotics as adjuvants for prevention and treatment of COVID-19: a bibliometric analysis and integrative review' [Trends in Food Science & Technology 120 (2022) 174–192]. Trends Food Sci Technol 121:156–160. https://doi.org/10.1016/j.tifs.2022.01.028

Xavier-Santos D, Padilha M, Fabiano GA, Vinderola G, Gomes Cruz A, Sivieri K, Costa Antunes AE (2022b) Evidences and perspectives of the use of probiotics, prebiotics, synbiotics, and postbiotics as adjuvants for prevention and treatment of COVID-19: a bibliometric analysis and systematic review. Trends Food Sci Technol 120:174–192. https://doi.org/10.1016/j.tifs.2021.12.033

Xie X, He Y, Li H, Yu D, Na L, Sun T, Zhang D, Shi X, Xia Y, Jiang T, Rong S, Yang S, Ma X, Xu G (2019) Effects of prebiotics on immunologic indicators and intestinal microbiota structure in perioperative colorectal cancer patients. Nutrition 61:132–142. https://doi.org/10.1016/j.nut.2018.10.038

Xu J, Ma Z, Li X, Liu L, Hu X (2020) A more pronounced effect of type III resistant starch: vs. type II resistant starch on ameliorating hyperlipidemia in high fat diet-fed mice is associated with its supramolecular structural characteristics. Food Funct 11(3):1982–1995. https://doi.org/10.1039/c9fo02025j

Yadav BS, Sharma A, Yadav RB (2010) Resistant starch content of conventionally boiled and pressure-cooked cereals, legumes and tubers. J Food Sci Technol 47(1):84–88. https://doi.org/10.1007/s13197-010-0020-6

Yang L, He QS, Corscadden K, Udenigwe CC (2015) The prospects of Jerusalem artichoke in functional food ingredients and bioenergy production. Biotechnol Rep 5(1):77–88. https://doi.org/10.1016/j.btre.2014.12.004

Yang L, Kang X, Dong W, Wang L, Liu S, Zhong X, Liu D (2022) Prebiotic properties of Ganoderma lucidum polysaccharides with special enrichment of Bacteroides ovatus and B. uniformis in vitro. J Funct Foods 92:105069. https://doi.org/10.1016/j.jff.2022.105069

Younes H, Coudray C, Bellanger J, Demigné C, Rayssiguier Y, Rémésy C (2001) Effects of two fermentable carbohydrates (inulin and resistant starch) and their combination on calcium and magnesium balance in rats. Br J Nutr 86(4):479–485. https://doi.org/10.1079/bjn2001430

Young P, Cash BD (2006) Probiotic use in irritable bowel syndrome. Curr Gastroenterol Rep 8(4): 321–326. https://doi.org/10.1007/s11894-006-0053-1

Yousefi M, Khorshidian N, Hosseini H (2018) An overview of the functionality of inulin in meat and poultry products. Nutr Food Sci 48(5):819–835. https://doi.org/10.1108/NFS-11-2017-0253

Zahid HF, Ranadheera S, Ajlouni S (2021) Utilization of mango, apple and banana fruit peels as prebiotics and functional ingredients. Agriculture 11:1–17. https://doi.org/10.3390/agriculture11070584

Zepeda-Hernández A, Garcia-Amezquita LE, Requena T, García-Cayuela T (2021) Probiotics, prebiotics, and synbiotics added to dairy products: uses and applications to manage type 2 diabetes. Food Res Int 142:110208. https://doi.org/10.1016/j.foodres.2021.110208

Chapter 6
Microbial Stabilizers in Food Processing

Bindu Devi, Sunita Devi, Diksha Sood, and Priyanka Arya

Abstract Food stabilizers are crucial food components that help the food retain its structure, improve its water-binding capacity, mouthfeel, thickening, consistency, and viscosity, among other things. They are obtained from diverse natural raw resources, viz. land and sea plants, animal's connective tissues, and microorganisms. Among these, microorganisms are receiving momentous attention as a stable and sustainable source of food stabilizers like xanthan gum, gellan, pullulan, and curdlan. Different kinds of microbes including bacteria, yeast, and actinomycetes have the capability to provide access to a larger range of food stabilizers. Also, genetically engineered microbes, viz. *Actinobacillus succinogenes, Basfia succiniciproducens, Mannheimia succiniciproducens, Escherichia coli*, etc., produce diverse stabilizers like amino and carboxylic acids as a part of their carbon's metabolism. This chapter delves in depth into the topic of food stabilizers, their sources, current endeavors to produce industrially relevant microbial food stabilizers along with current impediments.

Keywords Microorganisms · Microbial stabilizers · Metabolic pathways · Food processing · Food stabilizers

6.1 Introduction

Stabilizers are food additives that allow two or more immiscible constituents to disperse uniformly, improving food texture, color, and flavor stability, besides overall quality and acceptance. Natural gums have been reported to be utilized in food preparations since antiquity, but stabilizer's use in food preparations was only

B. Devi · S. Devi (✉) · D. Sood
Department of Basic Sciences, College of Forestry, Dr. YSP University of Horticulture & Forestry Nauni, Solan, Himachal Pradesh, India

P. Arya
Department of Food Science and Technology, College of Horticulture, Dr. YSP University of Horticulture & Forestry Nauni, Solan, Himachal Pradesh, India

A. K. Nadda, G. Goel (eds.), *Microbes for Natural Food Additives*, Microorganisms for Sustainability 38, https://doi.org/10.1007/978-981-19-5711-6_6

legalized around a half-century ago. The majority of stabilizers act by creating a molecular mesh network with three dimensions. They exert gelling effect, stiffening the medium and modifying its diffusion physiognomies. They are found in numerous foods including bread, dairy, fruits, vegetables, meat, and restructured meals. Stabilizers make it easier to manufacture fortified and composite foods with two immiscible constituents because they prevent food components from ghettoizing. Depending on where they emanate from and how do they interact with food, stabilizers have varying functional properties (Kumar et al. 2021). In a chemical sense, stabilizers are polymers having hydroxyl (OH) groups, which are accountable for linking with water (H_2O) molecules via hydrogen (H) bonds. They come in diversified shapes and sizes, including straight and branching with molecular weights varying from 105 to 106 kDa. Stabilizers take longer to dissolve in H_2O than smaller molecules as they are larger; shear to properly hydrate or a few necessitate elevated temperatures. They form highly viscous solutions even at minute concentrations upon dissolution. However, certain stabilizers form upon heating and/or refrigeration, or when cations are introduced to a solution.

6.2 Sources and Usages of Food Stabilizers

The major sources of stabilizers include trees, seeds, seaweeds, animals, and microorganisms. Seaweeds derived alginates plus carrageenans (from), tree seeds-derived locust bean plus guar gums, fruit pectin, and cotton-derived sodium carboxymethyl cellulose are classic examples of common stabilizers. However, xanthan, a bacteria-derived polysaccharide, and gelatin, an animal-derived polypeptide, are also employed occasionally. These biopolymers' nature is polymolecular and polydisperse as their topologies alter relying on the source and ambient conditions. Information pertaining to the frequently used food stabilizers, their sources, properties along with applications is portrayed in Fig. 6.1 and Table 6.1.

Stabilizers are employed to make functional foods like cereal plus legume-based beverages (soy, oat, and peanut milks), besides keeping fruit juice mixes and herbal extracts in juices stable. In the foodstuff, stabilizers are a good source of soluble fibers. They could likewise be employed to keep the fat's content besides maintaining meat and meat product's quality when these are frozen and thawed. Despite their functional attributes, stabilizers have plentiful health benefits, comprising cholesterol-lessening effects of cellulose, guar-gum (GrG) plus pectin; satiety-elevating effects of glucan plus alginate; anti-obesity effects of cellulose plus arabic gum; and antitumor property of carrageenan (Liang and Wang 2015).

Gelatin is among the most prevalently used commercial stabilizers in food, cosmetics, and biomedicine owing to its remarkable physiognomies as a gelling, binding, thickening, stabilizing, and film-creating agent. Until now, the utmost prevalent sources of commercial gelatins have been pig or cow bones and skins, owing to comparatively inexpensive budget of the finished gelatin product. Since demand-driven market for gelatin is dominated by beverage and food enterprises, its

Fig. 6.1 The most common food stabilizers and their sources. (Source: Kumar et al. 2021)

global demand is predicted to grow steadily due to continuous consumption trends in the pharmacological and nutraceutical industries (Ahmad et al. 2015). However, for certain groups of customers, utilization of animal-based gelatin in foodstuffs and pharmaceutical goods is circumscribed as animal sources do not match their religious requirements. Muslims and Jews are prohibited from ingesting pork-related goods; Hindus are not permitted to eat beef. However, vegans and vegetarians do not consume any animal products like meat. Consumers are particularly concerned about the spate of bovine spongiform encephalopathy, sometimes known as "mad cow disease," which happened during the 1980s and was allied with consuming contaminated cattle's gelatin (Maddern 2008). Consequently, food manufacturers are increasingly looking for alternatives to animal-based products. Although some substitutes match certain gelatin physiognomies, none of them yet meet all of gelatin's functional requirements.

Plant-derived macromolecules like starch, locust bean gum, alginate, carrageenans and guar gum have dominated the saleable market until now on account of their easy accessibility besides low purification costs. Notwithstanding, just a handful of these are exploited at commercial scale, microbe-derived polysaccharides have benefits over plant-derived macromolecules in respect of renewability, cost stability, and persistent and reproducible physicochemical features. However, polysaccharide-based gels do not possess the molecular backbone flexibility of animal-based gelatin, resulting in higher viscosities (Ahmad et al. 2015).

Table 6.1 The predominant food stabilizers, their physiognomies, and applications

S. no.	Natural stabilizers	Physiognomies	Applications in food sector	References
1	Tragacanth gum (TG)	• Expands fast in hot plus cold water	• Bakery toppings	Philp (2018)
		• Not dissolves in organic solvents like alcohol	• Confectionary cream	
		• Viscosity of 100–3500 centipoise	• Icings	
		• Acidic by nature	• Frostings	
		• Heating causes an irreversible drop in viscosity	• Sauces plus condiments	
			• Mayonnaise	
			• Jelly products	
			• Salad dressings	
2	Gum ghatti (GG)	• Protein-rich food, neutral flavor, and no odor	• Beverages	Philp (2018)
		• Ca^{2+} and Mg^{2+} salts are the most prevalent forms	• Oil in H_2O emulsions	
		• Restricted solubility in cold H_2O that increases above 90 °C	• As emulsifier in flavor oils	
		• Pseudoplastic behavior that isn't Newtonian		
		• Better emulsifying characteristics		
3	Arabic gum (AG)	• Easily liquefies in cold and hot H_2O up to a concentration of 50%	• Bakery	Elnour et al. (2018)
		• Newtonian behavior is observed with concentrations up to 40% while pseudoplastic behavior is observed at greater concentrations	• Confectionery	
		• Largest H_2O solubility (50% w/v) among the tree exudates and the lowest viscosity, allowing for high gum concentrations	• Beverage	
			• Microencapsulation	
4	Guar gum (GrG)	• Quickly hydrates in cold H_2O	• Bakery products	Mudgil et al. (2015)
		• Hydration for around 2 h is indispensable to get maximal viscosity	• Cake mixes	
		• With increasing shear rate, the viscosity in aqueous solutions decreases	• Ice cream	

(continued)

Table 6.1 (continued)

S. no.	Natural stabilizers	Physiognomies	Applications in food sector	References
		• Can be degraded by prolonged heating, resulting in low viscosity solutions	• Beverages	
		• The ideal temperature range for achieving maximum viscosity is 25–40 °C	• Cheese spreads	
		• Forms resilient and weak H-bonds in polar and nonpolar liquids, respectively	• Sauces	
		• Low hydration rates in the dissolved salts and other H_2O-binding compound's presence	• Salad dressings	
5	Tara gum (TG)	• Viscosity drops as the shear rate increases	• Bakery products	Wu et al. (2015)
		• Yields a viscous produce with a pleasant texture	• Starch formulations	
		• Viscosity reduces as the salt content rises	• Beverages	
		• pH of a 0.5% TG solution is 6.6 (at lower pHs, viscosity decreases, although it remains stable across a large pH range (3–11))	• Noodles	
		• Viscosity of TG solutions reduces as the temperature upsurges from 30 to 80 °C	• Edible coatings	
			• Yogurt	
			• Ice cream	
6	Karaya gum (KG)	• Among the gums, it has the least water solubility	• Whipped cream	Sahu et al. (2019)
		• Non-Newtonian pseudoplastic behavior	• Bakery products	
		• Displays better stability at low pH (acidic) range of 4.5–4.7	• Frozen dessert	
		• The pH value(s) greater than 8 may result in irreversible transformations	• Ice pops	
			• Meat products	
			• Sherbets	

(continued)

Table 6.1 (continued)

S. no.	Natural stabilizers	Physiognomies	Applications in food sector	References
7	Alginates, viz. propylene glycol alginates (PGA), alginic acid (AA), sodium (Na), calcium (Ca), ammonium (NH_4^+) and potassium (K), alginate	• Na, K, and NH_4^+ alginates are H_2O soluble	• Fruit and vegetables	Szekalska et al. (2016)
		• Ca and magnesium (Mg) alginates are H_2O insoluble	• Meat products	
		• Create gel at an acidic/low pH but high Ca concentration	• Extruded food products	
		• At 20 °C, a 1% w/v sodium alginate aqueous solution exhibits a dynamic viscosity of 20–400 mPas	• Desserts	
		• Prevent items from clinging to the wrapping paper	• Bakery toppings	
		• Lower syneresis	• Sauces	
		• Enhance product's shelf life	• Ice creams	
		• At low/acidic pH, alginates have restricted solubility	• Enzyme's encapsulation	
			• Flavors plus probiotics	
8	Pectin	• Pure water soluble	• Jams	Vanitha and Khan (2019)
		• Pseudoplastic behavior that isn't Newtonian	• Jellies	
		• High methoxyl (HM) pectins are hot H_2O soluble that requires a dispersant such as dextrose to avoid lumping while low methoxyl (LM) pectins form gels regardless of sugar concentration, although gelation requires a precise quantity of Ca plus other divalent ions	• Viscosity enhancer	
		• The pH range of 2–3.5 is mandatory to produce gels in HM while LM may create gels in the pH range of 2.0–6.0		
		• In the pH range 3–5, a high dose of Ca can promote cross-linking at a point where pectin precipitates and the gel gets ruined		

(continued)

Table 6.1 (continued)

S. no.	Natural stabilizers	Physiognomies	Applications in food sector	References
		• Thermal reversibility is a feature of HM gels, although LM gels display better stability at low/acidic pH but elevated temperatures		
9	Locust bean gum (LBG)	• Partially solvable in cold H_2O	• Beverages	Barak and Mudgil (2014)
		• For complete hydration, heat at 80 °C for 30 min	• Bakery products	
		• Maximal viscosity is attained in around 2 h of hydration	• Edible coatings	
		• The galactomannan chain undergoes oxido-reductive depolymerization at temperatures >80 °C, decreasing the solution's final viscosity	• Ice cream	
		• The ideal pH range (4–9) is acidic	• Noodles	
			• Low fat yogurt	
10	Carrageenan	• Hot H_2O soluble but Na salts dissolve in cold H_2O	• Frozen desserts	Tarande and Manriquez-Hernandez (2016)
		• Sulfate carrageenans create weak gels when temperature lowers down	• Cottage cheese	
		• Thermally reversible and has hysteresis	• Chocolate	
		• Stable at ambient (room) temperatures	• Whipped cream	
		• Heating over 5–20 °C, may dissolve the gel	• Yogurt	
		• Hydrolyzed at low/acidic pH (3.0), resulting in full loss of functionality	• Instant breakfasts	
			• Jellies	
			• Sauces	
			• Pet foods	
			• Syrups	
			• Tofu	
			• Milk shakes	

(continued)

Table 6.1 (continued)

S. no.	Natural stabilizers	Physiognomies	Applications in food sector	References
11	Agar-agar	• Unsolvable in cold H_2O but solvable in hot H_2O	• Cakes	Stanley (2006)
		• At 0.5–2% (low) concentrations, it creates a gel	• Toppings	
		• Up to 1.5%, a transparent solution is formed	• Icing	
		• Below 85 °C, the gel does not melt	• Bakery glazes	
		• Can contain a high amount of soluble substances like sugar without crystallization	• As clarifier for wine industry	
		• Gelation not require potassium or calcium		
		• Thickening and gelling qualities are excellent		
		• Prevents opacity and adhesive property loss		
12	Gelatin	• High melting and gelling points	• Ice creams	Echave et al. (2019)
		• Thermostatically reversible to a great degree	• Confectionary	
		• Collagen supply and tissue type affect its chemical characteristics	• Dairy	
		• Gelling strength of porcine skin gelatin (PSG) is greater than that of bovine skin gelatin (BSG)	• Edible packaging films	
		• BSG has stronger foaming qualities than PSG	• Meat products	
		• The stability of fish gelatin is lesser than human gelatin		
13	Milk proteins	• Exposure to heat under 95 °C for some minutes generates reversible alterations in casein micelles and salt distribution	• Confectionary products	Jeewanthi et al. (2015)
		• Heat treatment beyond 120 °C causes permanent modifications	• Bakery products	
		• Whey proteins may be subjected to hydrolysis under controlled environment	• Pastries	

(continued)

Table 6.1 (continued)

S. no.	Natural stabilizers	Physiognomies	Applications in food sector	References
		• Foam stability, creaming ability, and emulsifying activity and stability may all be improved	• Whipped toppings	
		• Hydrolysis ensued in a modest reduction in gelling ability	• Yogurts and infant formulae	
		• At pH 4.0, hydrolysates of whey protein are less soluble, whereas at pH 6–10, they are most soluble	• Ice creams with reduced fat content	
14	Dextran	• Soluble in solvents like H_2O, glycerol, glycol, and ethylene	• Confectionary	Kothari et al. (2014)
		• Insoluble in ethanol, methanol, isopropanol, 2-propanone, and acetone	• Bakery	
		• Linear dextrans exhibit great H_2O solubility plus Newtonian behavior while branched dextrans display pseudoplastic behavior	• Dairy	
		• Viscosity remains unaffected by alterations in salt concentrations or pH	• Prebiotics	
		• Up to 300 °C, they have exceptional thermal stability	• Frozen products	
15	Xanthan gum (XG)	• Solvable in both hot and cold H_2O	• Baked goods	Lopes et al. (2015)
		• The XG's viscosity falls with raised shear rate	• Beverages	
		• Viscosity upsurges with elevated temperature (from 40 to 60 °C), whereas it decreases above 60 °C	• Frozen foods	
		• Over an extensive pH range (1–13), the viscosity stays constant	• Dairy products	
			• Syrups	
			• Salad dressing	
16	Chitin		• Coating of foods	

(continued)

Table 6.1 (continued)

S. no.	Natural stabilizers	Physiognomies	Applications in food sector	References
		• Unsolvable in organic and inorganic solvents		Roy et al. (2017)
		• Chitin's solubility may be improved by transforming it into chitosan, which is easily solvable in dilute acidic medium (pH < 6.5)		
		• At low/acidic pH, the ionization of chitosan molecules rises		
		• With increasing ionic strength, the chitosan solution's intrinsic viscosity decreases		
17	Gellan gum (GG)	• Unsolvable in cold H_2O, but stirring easily disperses it	• Gummy candy	Osmałek et al. (2014)
		• GG gels exhibit rheological behavior (thermo reversible) and have alike characteristics at 20 °C as they show at 90 °C	• Starch jellies	
		• Quick set time and high gel strength	• Marshmallow	
		• Gel's stiffness diminished when the pH rises	• Bakery fillings	
		• Provides a magnificent clarity		

Many research have discovered that microbe-derived polysaccharides have the prospective to act as gelling, viscosifying, emulsifying, or stabilizing agents in different foodstuffs and pharmaceutical goods as a substitute to natural components obtained from animals and plants. Numerous microbe-derived polysaccharides are renowned for displaying anticancer, antioxidant, antibacterial, antiulcer, and cholesterol reducing properties. Consequently, microbe-derived polysaccharides have attracted growing interest since the unearthing of gellan, xanthan, and dextran owing to their unique functions, physicochemical characteristics, reproducibility, stability, and production cost (Falconer et al. 2011). Furthermore, altered forms of cellulose and starch are correspondingly employed as stabilizers. The properties along with applications of most prevalently used stabilizers are depicted in Table 6.1.

6.3 Properties and Selection Criteria of a Good Stabilizer

Food manufacturers are continuously on the lookout for the ideal food stabilizer to help them increase their product's acceptability and outspread their market. A good stabilizer should be colorless, odorless, and have a great moisture retention rate. To acquire the desired outcomes in food items, stabilizers can be engaged singly or in amalgam. An ideal food-grade stabilizer must meet the following criteria:

1. Should be nontoxic.
2. Should be dispersed freely in the mix.
3. Should not produce excessive viscosity, separation, or froth in the mix.
4. Filters and strainers must not be bunged.
5. Should not provide off-savor to the food stuff.
6. At usage time, availability and pricing are critical.

6.4 Selection Criteria

The following parameters should be borne in mind before deciding which stabilizer to employ:

- Solution clarity
- Temperature solubility
- Viscosity or gel properties required
- Dispersion issues
- Temperature and other processing settings
- Hydration rate
- Suspension ability
- Natural vs. synthetic
- Particle size required
- Capability to make proteins stable at lower pH
- Acid stability
- Availability and price at the usage level

6.5 Microbial Food Stabilizers

Biopolymers or food stabilizers, which are environmentally and human friendly, have been envisaged as biotechnological products in an array of industries, including food, agriculture, and health, in recent times. Food items require long-chain polymers having high molecular masses that dissolve/disperse in H_2O to offer gelling/or thickening properties. Food polymers are betrothed in secondary ejects such as emulsification, stability, particle suspension, crystallization control, syneresis inhibition (water escape from processed meals), film production, and encapsulation. The predominantly utilized biothickeners in the food trade nowadays are plant-derived polysaccharides (e.g., alginate, starch, carrageenan, pectin, locust bean, and guar gums) or seaweeds. Animal-derived proteinaceous hydrocolloids like casein and gelatin are also employed. These polymers' functional attributes in food are revealed by minute structural characteristics. These polysaccharides, however, unable to be accessible at all times in the requisite grade, and their rheological capabilities may vary slightly. Most plant sugars are chemically altered with the eventual objective of improving their rheological physiognomies and structure. Thus, their use is very restricted (Freitas et al. 2021).

Biothickeners/stabilizers made from microbial (microbe-synthesized) exopolysaccharides (EPS) denote a different form of biothickeners/stabilizers. Extracellular polysaccharides adhering to the cell's surface as capsules or discharged into its exterior environment as slime are adverted to as microbial exopolysaccharides. Capsules or slimes are the terminology used to characterize them. EPS is abundant in both bacteria and microalgae, though it is scarce in fungi including yeasts. Microbial cells are presumed to use EPS to defend themselves from phagocytosis, desiccation, antibiotics, phage attack, and toxic chemicals (e.g., toxic metallic ions, sulfur dioxide, ethanol), protozoan predation, osmotic stress, adherence to solid substratum/surfaces, and biofilm development, besides recognizing themselves in their natural settings (e.g., via cohering to a lectin). Since a good bit of slime-producing bacteria are unable to catabolize the EPS they form, EPS is unlikely to function as a food reservoir. Both capsular EPS and O-antigen lipopolysaccharides (LPS) are accountable for the immunogenic response in pathogenic bacteria like *Streptococcus pneumoniae*, *S. agalactiae*, etc. (Sousa et al. 2013).

Dextrans, gellan, xanthan, pullulan, bacterial alginates, and yeast glucans are a handful of classic examples of microbial exopolysaccharides utilized in industry. Novel biopolymers, particularly microbial polysaccharides, unwrap new avenues for the replacement/substitution of either standard product or filling the gap in these biopolymer's market accessibility, concerning higher stability and rheological properties.

Microbial polysaccharides now constitute a modest portion of the present-day biopolymer's market. However, two limitations limit their applicability: the expensive cost of collecting microbial EPS, which demands a complete comprehension of their production besides suitable bioprocess technology. So far, just a few microbial stabilizers have been marketed at industrial level, including *Xanthomonas campestris*-derived xanthan gum, dextran from *Leuconostoc mesenteroides*, *Sphingomonas elodea*-derived gellan gum, and succinoglycan (*Rhizobium* sp.).

Xanthan is a microbe-derived EPS which has been legalized/approved for usage in food industries on account of its inexpensive production cost and exclusive rheological qualities. However, *Xanthomonas campestris* phytopathogen, which is not generally recognized as safe (GRAS), produces huge levels of it. Similarly, gellan derived from *Sphingomonas elodea* (phytopathogenic bacterium) has already been available on the market. Many of the aforementioned problems could be addressed by GRAS, or food-grade microbes like lactic acid bacteria (LAB) that can produce large volumes of EPS. Table 6.2 portrays the instances of those bacterial polysaccharides that exhibit commercial potential.

Table 6.2 Commercially viable microbial sources of polysaccharides

Sr. no.	Microbial sources	Polysaccharides	References
1	*Xanthomonas*	Xanthan gum	Ahmad et al. (2015)
2	*Pseudomonas putida*	Alginate	Celik et al. (2008)
3	*Leuconostoc mesenteroides*	Dextran	Falconer et al. (2011)
4	*Zymomonas mobilis*	Levan	Ahmad et al. (2015)
5	*Leuconostoc dextranicum*	Glucan	Majumder and Goyal (2009)
6	*Gluconacetobacter xylinum*	Cellulose	Huang et al. (2010)
7	*Sphingomonas elodea*	Gellan	Ahmad et al. (2015)
8	*Agrobacterium radiobacter*	Succinoglycan	Ahmad et al. (2015)
9	*Aureobasidium pullulans*	Pullulan	Andhare et al. (2014)
10	*Alcaligenes faecalis*	Curdlan	Andhare et al. (2014)
11	*Acetobacter xylinum*	Acetan	Jindal and Khattar (2018)
12	*Lactobacillus hilgardii*	Kefiran	Jindal and Khattar (2018)
13	*Alcaligenes* spp.	Welan	Andhare et al. (2014)
14	*Acinetobacter calcoaceticus*	Emulsan	Jindal and Khattar (2018)

6.5.1 *Xanthan*

This bacterial exopolysaccharide is produced by a bacterial pathogen, *Xanthomonas campestris*, as xanthan gum (XG) through aerobic and submerged fermentation utilizing glucose as the principal carbohydrate's source. The xanthan is extracted, refined, dried, and ground into a white colored powder. XG is a linear plus anionic hydrocolloid with (1 → 4) allied backbone of α-D-glucose molecules. A glucuronic acid moiety is coupled to a mannose unit present at the terminal end via 1 → 4 linkage while to a second mannose moiety via 1 → 2 linkage in a side unit, forming a trisaccharide.

Xanthan is generally stable at wide-ranging pH values (2–12). The solution's ionic strength has a momentous influence on how xanthan gum works. It works as a thickening/gelling agent that functions well in a diverse array of applications. It possesses pseudoplastic rheological physiognomies, i.e., viscosity reduced with augmented shear. It works better as a stabilizer in sherbet, ice cream, H_2O ices, and ice milk when combined with LBG and/or guar gum. The usage of sodium alginate with XG as a milkshake stabilizer has already been documented (Huang et al. 2010).

6.5.1.1 Chemical Composition and Structure

Xanthan is basically a polysaccharide (long-chained), which is comprised of a significant number of side chains of trisaccharide moieties with 3:3:2 molecular ratio. Its building block units are glucose, mannose, and glucuronic acid. Typically, xanthan possesses a molecular weight (MW) of around 2000 kDa (Jindal and Khattar 2018).

6.5.1.2 Physiological Characteristics of Xanthan

Xanthan is a H_2O-solvable hydrocolloid that dissolves quickly and easily when stockpiled at ambient (room) temperatures. Constant mixing and swirling reduces xanthan's hydration time. Xanthan polymers possessing long chains are easy to dissolve/disperse, but they take longer to get hydrated. Ionic strength, generally, determines hydration. The hydration process is slow when great ionic strength is there. As temperature elevates, the otherwise stiff, well-organized state transforms into a supple, disorderly state and the solution experiences conformational shift. Xanthan solutions display viscoelastic behavior, ensuing in a fragile gel which suspends liquid food stuffs, viz. sauces, dressings, and cakes prior to baking. The widespread pH values (3–10) have little impact on xanthan solution's behavior. It is documented to be more stable and resistant to proteases, cellulases, and amylases in comparison to other thickeners (Wustenberg 2014).

6.5.1.3 Applications

Xanthan gum is mostly utilized in dressings and sauces. It aids gel formation but reduces syneresis in dairy desserts (Jindal and Khattar 2018).

6.5.1.4 Legislation and Toxicity-Related Issues

The FDA authorized the usage of XG in the USA in 1969 in foodstuffs. In 1974, Europe allowed its usage as a culinary additive. All animal trials, including toxicological and biochemical tests, have advocated its use safe (Imeson 2010). The Food Safety and Standards Authority of India (FSSAI) has recommended the usage of XG in whip-tops (non-dairy) and Bakery mixes @ maximum 0.5% w/v in India (FSSAI 2006).

6.5.2 Gellan Gum (GG)

This fermentation polysaccharide is produced by *Sphingomonas elodea* ATCC 31461, hitherto called as *Pseudomonas elodea*. Gellan gum has a linear chemical structure and is comprised of repeating glucuronic acid rhamnose, and glucose units. When gellan solutions cool, the polysaccharide strands are transformed into double helices, resulting in fragile gel structures. In the existence of suitable cations (Na^+ or Ca^{++}), the double helices are formed (cation-mediated aggregation), ensuing the formation of resilient gel networks (Wang et al. 2006).

6.5.2.1 Structure and Chemical Composition

Gellan is a negatively charged (anionic) heteropolysaccharide possessing a linear structure, comprised of D-glucose, D-glucuronic acid, and L-rhamnose (molecular ratio: 1.5:1:1) units. It is composed of reiterating units of tetrasaccharide moieties in which glucuronic acid, glucose plus glucose (β-1 → 4 allied), and rhamnose (α-1 → 3 linked) are linked. To put it another way, two β-1,4,D-glucose units, one α-1,3 L-rhamnose unit, and one β-1,4,D-glucuronic acid unit make up the tetrasaccharide. GG creates gels at very tiny concentrations, i.e., 0.1%. They come in two varieties: (a) low acyl, which produces hard and fragile gels, and (b) high acyl that produces soft, supple, and nonfragile gels. Gellan gum exhibits an ionotropic gelation, akin to carrageenan or alginate polysaccharides. GG forms coiled structure at raised temperatures, but once the temperature drops, it transforms into a double-helix structure. These helices are self-assembled to produce junction zones (orientated bundles) accompanied by coupling with untwisted regions in the polysaccharide strand, culminating into a gel's development (Quinn et al. 1993; Oliveira et al. 2010).

6.5.2.2 Physical Properties

The physical and functional characteristics of GG are influenced by acyl groups' presence or absence. GG dissolves quickly in both hot and cold H_2O. It creates a mesh-like structure termed demoldable gel when its concentration is more in the solution. On the contrary, when it occurs in minute concentrations, the molecules cluster together; however the solution remains very fluidic, hence creating a fluid gel. Gellan gum could likewise be employed to amend the stability, rheology, and thermal properties of solutions by uniting it with another non-gelling and gelling hydrocolloids (Coviello et al. 2007). Gellan and xanthan or carboxymethyl cellulose are used to form soft gels (Stephen et al. 2006).

6.5.2.3 Applications

GG is prevalently utilized as a binder, thickener, and stabilizer in an array of culinary applications. Additionally, it is utilized to keep water-based gels like sipping jellies and desserts stable. Gellan can even be utilized in vegan diets to substitute gelatin in yogurt and sour cream like dairy products. Gellan is documented to be excellent for speeding up the setting of starch confections and keeping sweets from sticking/adhering together in hot weather (Mariod and Fadul 2013).

6.5.2.4 Concerns About Legislation and Toxicity

In 1988, Japan became the first country to legalize the usage of GG (Bajaj et al. 2007). In 1997, GG was approved for usage in foods, cosmetics, medicines, and nonfood goods in 25 countries all through the world, including Japan, the USA, Canada, South Africa, Australia, South America, and Southeast Asia. It had FDA approval in the USA; nonetheless it is categorized as a "food additive" (E 418) in Europe (Venugopal 2011).

6.5.3 Curdlan

Curdlan (E424) is a neutral, linear, and unbranched β-D-glucan polysaccharide comprised of β-1,3-allied repeating glucose moieties. It is named as linear polymer on account of its capability to "curdle" when imperiled to heat. Curdlan is mainly obtained from bacteria, for example *Agrobacterium* biovar.1, *Alcaligenes faecalis* strain 10C3, a few *Rhizobium* spp., and Gram-negative *Cellulomonas* spp., especially *C. flavigena* KU (McIntosh et al. 2005). At ambient (room) temperature, it is entirely unsolvable in H_2O besides maximum organic solvents. However, when heated, it produces two forms of heat-induced gels: "high-set gels" and "low-set gels" (Lee 2002). Curdlan gels possess physiognomies which are akin to gelatine's suppleness and agar's fragility. In an assorted range of culinary preparations, such as surimi products, processed meat, vegetarian meals, sauces, and other functional foodstuffs, it is also utilized as a texturant processing aid, stabilizer, texture adjuster, or thickener (Spicer et al. 1999).

6.5.3.1 Structure and Chemical Composition

It is a succinoglucan comprised of linear (β-1 → 3) glucans (Fig. 6.2) that are partly esterified with succinic acid (Wustenberg 2014). Curdlan is mostly unsolvable in H_2O under 54 °C temperature. When solution's temperature rises beyond this level,

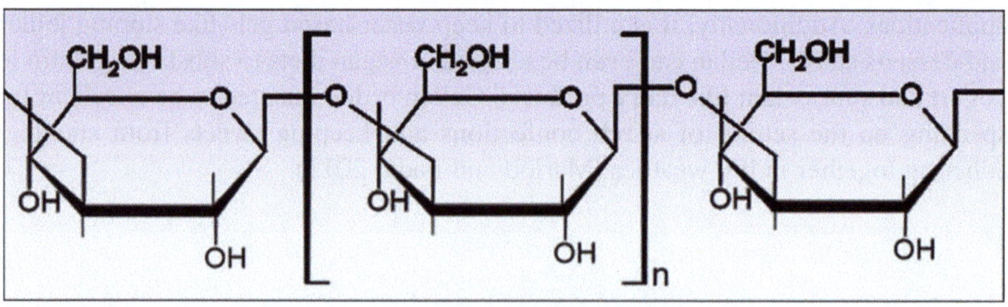

Fig. 6.2 Curdlan's chemical structure. (Source: Jin et al. 2006)

it expands and forms a sturdy supple gel. Curdlan is solvable in those aqueous solutions which are alkaline in nature (Saha and Bhattacharya 2010). Curdlan possesses three –OH groups (chemically active), one at carbon 6 position (primary –OH group) while two at carbons 2 and 4 positions (secondary –OH groups). Being a primary OH group, O-6 is the core favorable location for substantial changes which is particularly expedient in regioselective processes (Nasrollahzadeh et al. 2021).

6.5.3.2 Physical Properties

During the course of heating, curdlan transforms into an irreversible gel. With the rise in temperature, the gels' strength increases. In water, curdlan forms unsolvable films, which are eatable, resistant to oxygen, and biodegradable too (Ghanbarzadeh and Almasi 2013).

6.5.3.3 Applications

Curdlan performs the function of a gelling/thickening agent both in food and pharmaceutical sectors. It is utilized in making jellies, sweets, and confectionary instead of gelatin and agar; also utilized as a binder and thickener in dietetic meals (e.g., salad dressings, pasta, and desserts), besides as an edible coating for enhancing food's shelf life (Nishinari et al. 2021).

6.5.3.4 Legislation and Toxicity-Related Issues

The US Food and Drug Administration (FDA) has approved the usage of curdlan as a food ingredient. Curdlan has already been granted a food license in Japan (Venugopal 2011).

6.5.4 Pollulan

Pollulan was first produced in 1976 in Japan at commercial scale. Pollulan is a kind of extracellular glucan secreted by *Aureobasidium pullulans* (fungal strain), sometimes acknowledged as "black yeast." Microbial cells are filtered out of the soup to purify pollulan. The filtrate is decolorized with activated charcoal, and excess salts and proteins are removed with ion exchange chromatography. The concentrated solution is dried and crushed so that a fine powder can be obtained (Choi et al. 2014).

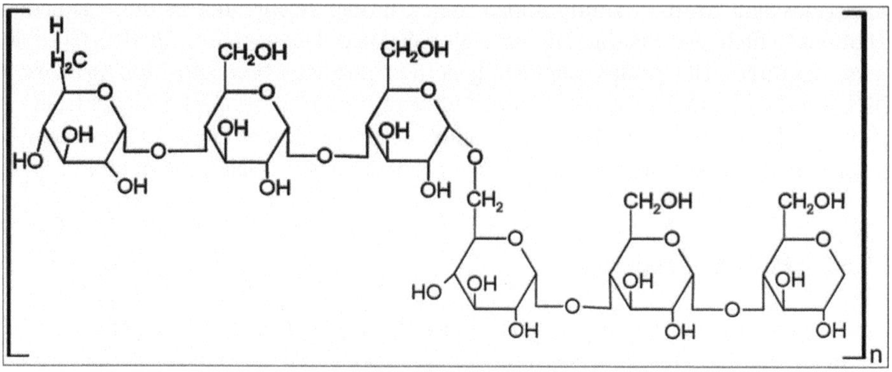

Fig. 6.3 Pollulan's chemical structure. (Source: Jindal and Khattar 2018)

6.5.4.1 Structure and Chemical Composition

Pollulan is a water dispersible linear homopolysaccharide primarily composed of α-(1 → 6) allied maltotriose units, with subsequent maltotriose units joined by α-(1 → 6) glycosidic linkages (Fig. 6.3). The pollulanase enzyme hydrolyzes α-(1 → 6) bonds in pollulan and converts them to maltotriose, as identified later. Nevertheless, it is resistant to glucose oxidase, amylases, invertase, fructosyltransferase, glucosidase, and a handful of proteolytic enzymes (Nasrollahzadeh et al. 2021). The pollulan's molecular mass varies from 10 to 400 kDa (Navard 2013).

6.5.4.2 Physical Properties

Pollulan is a white powder without any odor and taste. Its solution is resilient to heat and stable across an extensive pH range. When dissolve in water, it functions as an adhesive and sticks to wood better than modified cellulose or cornstarch. Pollulan produces an edible, translucent, oil-impermeable and water-soluble film in a little while (Utami et al. 2014; Ferreira et al. 2016).

6.5.4.3 Applications

Pollulan is mostly utilized as an edible, filler and tasteless polymer in the food sector. It is, generally, utilized in icings, instant beverages, soy sauces, creams, confectionery, sweets, and other products. Pollulan exhibits some properties like (a) a superior oxygen barrier, (b) good capability to retain moisture, and (c) inhibits fungus growth, making it an efficacious food preservative. Its most common application is as a calorie free/low food additive that could be substituted for gelatin in coatings.

Pollulan is a polymeric system used to form hydrogels and nanogels. These systems could be engaged to fabricate green smart materials, biosensors, biomimetics, artificial muscles and chemical separations, among other items. In electronics and pharmaceuticals, it is employed as a coating and a fiber and film-forming agent in tissue engineering, drug and gene delivery systems, because of its virtuous physicochemical qualities (Nasrollahzadeh et al. 2019). Pollulan's use as a carrier/transporter for transmucosal medicine delivery systems in the nose, mouth, and lungs has been documented as well (Narayanaswamy et al. 2016).

6.5.4.4 Legislation and Toxicity-Related Issues

The US FDA declared pollulan's usage safe, in 2002. Recently, the European Union (EU) has approved its usage as a food additive (E 1204) in films, tablets, and capsules under Directive—2006/52/EC. Now, the usage of pollulan is also legal in the food sector in various South American and Asian countries like Russia (Venugopal 2011).

6.5.5 Dextran

Generally, lactic acid bacteria (LAB) including *Leuconostoc mesenteroides* along with a few strains of *Gluconobacter oxydans* are employed for dextran production via fermentation. The former strains utilize dextransucrase enzyme to transform sucrose into dextran whereas maltodextrins are converted to dextran by the dextran-dextrinase enzyme by the latter strains (Naessens et al. 2005). During fermentation, microorganisms are eliminated from the cultural media. Organic solvents like methanol, ethanol, acetone, 2-propanol, and others are utilized to precipitate the solubilized dextrans. Reprecipitation is employed to remove entrapped contaminants. A fine white colored powder is the finished/final product (Wustenberg 2014).

6.5.5.1 Chemical Composition

Dextran is basically a generic term used to denote those polymers, which are comprised of D-glucose moieties linked by α-(1 $\rightarrow$ 6) links with flexible numbers of side branches attached by α-(1 $\rightarrow$ 2; 1 $\rightarrow$ 3 or 1 $\rightarrow$ 4) linkages to the primary chains. The dextran's molecular mass can reach up to 440 MDa (Zarour et al. 2017) and is classified into two kinds based on their chains' length: dextrans possessing a MW of $>$40 kDa (Heinze et al. 2006) and oligodextrans possessing a MW of $<$40 kDa (Díaz-Montes 2021).

6.5.5.2 Applications

Dextran's chief applications are in the food trade, where it is employed in confectionery and baking to improve/enhance the flavor, texture, along with consistency of sweets, ice creams, breads, jellies, and flours on account of its moisturizing, stabilizing, and preserving attributes. It is also utilized to prevent oxidation in vegetable, meat, and cheese; also it enhances aroma, texture, and flavor (Aman et al. 2012; Wolter et al. 2014). Additionally, dextran has also been recommended for usage as potential prebiotics and biodegradable film-forming agents or coatings (Sarbini et al. 2013).

Dextran may be utilized in multitudinous applications. For instance, high MW (40–70 kDa) dextran has been reported to be utilized in medications as a cryopreservative for organs and vaccines, besides an anticoagulant, extender, osmotic agent, antithrombotic and intravenous plasma lubricant (Bhavani and Nisha 2010). Also it is utilized as a moisturizing and/or thickening ingredient in cosmetics, and its reductive properties permit its usage as a wrinkle reducer ingredient (anti-aging agent).

6.5.5.3 Legislation and Toxicity-Related Issues

Dextran may be consumed as an instant energy source as it is simply absorbed in the digestive system of human beings. Its removal is easy by the kidneys because it expands blood plasma. Therefore, dextran is presently authorized in several countries for usage in an extensive array of foodstuffs. However, dextran is not endorsed to be consumed/eaten on a daily/regular basis (Jindal and Khattar 2018).

6.5.6 Scleroglucan

The term scleroglucan denotes a form of extracellular glucans produced by phytopathogenic fungi, particularly those belonging to the *Sclerotium* genus, viz. *Sclerotium rolfsii, S. glucanicum*, *Botrytis cinerea, Schizophyllum commune*, and *Epicoccum nigrum*. Scleroglucan is hypothesized to be produced by these phytopathogens to attach to the plant's surface. Furthermore, the cells that secrete scleroglucan hydrolyze it to glucose, suggesting that it functions as a carbohydrate store. The commercial product's brand name is Scleroglucan, although it is also acknowledged as clearogel, actigum, polytetran, polytran FS, and sclerogum (Jain et al. 2007; Schmid et al. 2011).

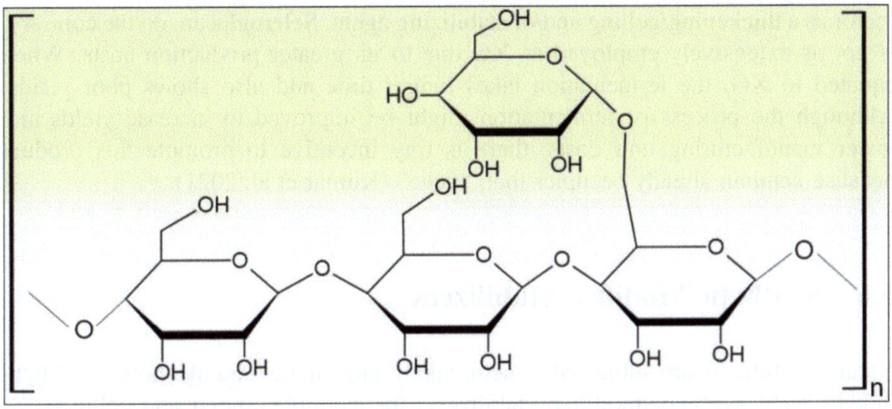

Fig. 6.4 Scleroglucan's chemical structure. (Source: Kırtel et al. 2017)

6.5.6.1 Chemical Composition

Scleroglucan is a homopolysaccharide (H_2O-dispersible plus nonionic) comprised of a linear strand of β-D-(1 → 6)-glucopyranosyl and β-D-(1 → 3)-glucopyranosyl groups (Fig. 6.4). The length and frequency of side chains differ among scleroglucan-producing microbial species. The MW of scleroglucan is ~2 kDa. Its structure was first revealed by periodic oxidation analysis, which was subsequently verified by 13C nuclear magnetic resonance (NMR) and methylated sugar analysis (Survase et al. 2007).

6.5.6.2 Physical Properties

It dissolves quickly in hot and cold H_2O. Neither acids nor alkalis affect the scleroglucan solution throughout an extensive pH range of 2.5–12.0 since it is nonionic in nature. Also ions do not influence the scleroglucan solution's viscosity besides being pseudoplastic and thermostable, with a great yield. Scleroglucan is a foam stabilizer with excellent emulsification abilities. On account of its strong rheological characteristics and durability over a wide-ranging pH, salinities, and temperatures, scleroglucan is ideal for a varied array of applications. It is utilized to protect sauces and ice creams from getting too soft. Several patents describe ways to improve/enhance the quality of heat-treated or frozen meals, viz. Japanese cakes, steaming dishes, bakery items, and rice crackers (Zhang et al. 2013).

6.5.6.3 Applications

Scleroglucan shares most of the alike characteristics with xanthan and is at least effective with regard to functional properties. It may be widely utilized in the food sector as a thickening/gelling and/or stabilizing agent. Scleroglucan, on the contrary, is not as extensively employed as XG due to its greater production costs. When equated to XG, the fermentation takes longer time and also shows poor yields. Although the process of fermentation might be improved to increase yields and lower manufacturing unit costs, there is tiny incentive to promote this product because xanthan already occupies that "niche" (Kumar et al. 2021).

6.6 Synthetic/Modified Stabilizers

Natural stabilizers are subjected to acid, alkali, enzymatic, and alcoholic modification to yield synthetic/modified stabilizers; for example, starch and cellulose as described in the following sections.

6.6.1 Modified Starch

A naturally derived polymer from agricultural sources is starch. It occupies second place among utmost prevailing natural polymers that exist on the globe, after cellulose. Wheat, tapioca, rice, corn, and potato are only a handful of the food industry's most popular cereal grains and tubers. Starch contains two forms of homopolysaccharides—amylose and amylopectin. The former one polymer is extensively branched encompassing around 1,000,000 glucose moieties united by α-1,4 bonds in the linear strand while through α-1,6 glycosidic linkages in the branched strands. Amylopectin is an extremely branched polymer having approximately 500–20,000-D-glucose moieties linked via α-1,4 bond in the linear strand while through α-1,6 glycosidic linkages in the branched strands. Amylose (20–25%) and amylopectin (75–80%) make up the major portion of starches. Other amylose-containing starches have 80% amylose content (Li et al. 2019).

6.6.1.1 Modification

Starches are transformed by physical, chemical, enzymatic, or genetic techniques to change their inherent characteristics and improve their stabilizing qualities. Other modification procedures like dry heating, annealing, elevated moisture treatment, gelatinization, profound freezing/thawing, ultrasonication, osmotic pressure treatment etc. employed for modified starch/cellulose's manufacturing are depicted in Table 6.3. Physical approaches lower down the starch's viscosity by converting its granular structure into a tiny crystalline structure, allowing it to dissolve in cold/ice-cold H_2O (Li et al. 2019). Conversely, starches modified via physical means have restricted stabilizing properties. The chemical-mediated starch's modification brings about the insertion of a functional group into its chain, ensuing different

Table 6.3 Properties and production mechanisms of major modified starches and celluloses

S. no.	Modified cellulose/starch	Production mechanism	Physiognomies	Applications in food sector	References
1	Dextrins	Ortho-phosphoric or hydrochloric acids (HCl) are used to dry or roast native starch	Reduced gel viscosity, enhanced stability, improved cold water solubility, and emulsification physiognomies	In confectionary/candies coating, fat substitute, encapsulating flavor	Chiu and Solarek (2009)
		Drying is employed for making dextrin roasted starch			
2	Bleached starch	Bleaching substances viz. hydrogen peroxide, peracetic acid, sodium hypochlorite (NaOCl), sodium chlorite, sulfur dioxide (SO_2), alternate allowed forms of sulfites, ammonium persulfate, and potassium permanganate ($KMnO_4$) are utilized to bleach native starch	Reduced viscosity, improved granule whitening, high paste clarity, stability at lower temperature, and improved adhesion	Soups and sweets candies' coating and as a binder in confectionary	FAO, WHO (2020)
3	Acid modified starch	Treating slurry with either diluted HCl/orthophosphoric acids, or sulfuric acid (H_2SO_4) or via native starch's suspension	Swelling power and solubility are improved, hot-paste viscosity is reduced, and textural features are improved	Gum, pastilles, jellies, tablet binders, candy making	Singh and Ali (2008)
4	Oxidized starch	Sodium hypochlorite treatment of food starch	Reduced viscosity, enhanced granule whitening, high paste clarity, better stability at lower temperature, and better adhesion	Breaded foods, binder in confectionery and meat, in fried food's coating and improving dairy product's texture	Korma et al. (2016)
5	Distarch phosphate	Cross-linking of two –OH groups on nearby anhydrous glucose moieties by the alike phosphate group	Displays stability at raised temperature, shear, and lower pH along with imparting increased hardness to the swollen starch granule, providing viscosity and resistance to syneresis process during storage besides improving texture	Thickener and stabilizer in sauces and soups	Krempel et al. (2019)

(continued)

Table 6.3 (continued)

S. no.	Modified cellulose/starch	Production mechanism	Physiognomies	Applications in food sector	References
6	Alkaline modified starch	Reacting native starch solution or slurry with sodium/potassium hydroxide	Increased viscosity, unsolvable in cold H_2O and ethanol, In hot H_2O, forms viscous colloidal solutions	In starchy food stuffs with varied digestibility	Qin et al. (2019)
7	Monostarch phosphate	Esterification of a phosphate to a single –OH group on a single anhydrous glucose moiety	Increased viscosity; improved luster and transparency	Flavor encapsulation, tomato ketchup/paste, soups, and sauces	Krempel et al. (2019)
8	Enzyme-treated starch	Treating native starch suspension with amylolytic enzymes (food grade), viz. α-amylase, glucoamylase, β-amylase, isoamylase, and pollulanase	Increased resistance, starch concentration, and an alteration in starch structure	Coating of ready-to-eat/breakfast cereals	Luckett and Wang (2012)
9	Methylcellulose (MC)	Synthesized via cellulose's etherification that involves reaction among alkali, cellulose, and chloroethane/iodoethanesulfites	White-colored powder, which is odorless, tasteless, solvable in H_2O, and organic solvents; in cold H_2O, creates a gel layer; upsurge in temperature causes an augmentation in viscosity that eventually ensued thermoreversible gels' formation	Bakery products (gluten free), mayonnaise, and cold drinks	Nasatto et al. (2015)
10	Carboxymethyl cellulose (CMC)	Synthesized by reacting cellulose and monochloroacetic acid in alkali's presence	An odorless and tasteless white/ cream white powder that exhibits varied viscosity and better heat stability	Ice cream, beverages, and bakery products	Jia et al. (2016)
11	Microcrystalline cellulose (MCC)	Synthesized by reacting cellulose and methyl chloride at high pressure, using alkali as a catalyst	Handling is difficult so entails homogenization with co-agents	Bakery products and ice creams with low fat content	Philp (2018)
12	Hydroxypropyl methylcellulose (HPMC)	Synthesized by reacting cellulose and propylene oxide using alkali as a catalyst and applying high pressure	When dispersed in H_2O, a viscous, opalescent, and colloidal solution is formed; Solvable polar solvents like H_2O but unsolvable in hot H_2O, diethyl ether, acetone, and alcohol	Fried foods	Philp (2018)

modifications in starch characteristics. Chemical-mediated starch's modification could be accomplished through oxidation, alkali/acid addition, bleaching, esterification, etherification, cross-linking, or a blend of all these processes. The hydroxyl (OH) groups situated at 2, 3, and 6 carbon atoms of glucose's pyranose structure do not have glycosidic connections and are thus available for replacement. During the enzymatic transformation, the glycosidic linkages are broken. Dextrins, enzyme-treated and acid/alkali modified starches are synthesized by hydrolysis. The bond cleaves when water molecules are added across it, resulting in cleavage products having an alcohol (hydroxyl group) functionality. When starch interacts with NaClO (sodium hypochlorite), oxidized starch is formed. It produces a whiter starch with improved water absorption, viscosity, and solubility properties (Egharevba 2019). Cross-linking occurs when starch interacts with a chemical, capable of forming bonds with several –OH groups in the molecule. Cross-linking improves freezing-thawing stability by lowering starch swelling power, enhancing solubility, improving shear and heat stability, and lowering starch swelling power. The hydrophilic –OH groups at 2, 3, and 6 carbon positions of starch are changed to hydrophobic acetyl groups when starch undergoes reaction with acetic anhydride. Acetate is generated when the acetic anhydride's acetyl groups esterify the –OH group of glucose. Acetylation improves starch granule solubility, swelling strength besides reducing gelatinization temperatures. When starch undergoes esterification with succinic anhydride, it is referred to be succinylated. Consequently, the insertion of more number of lipophilic groups takes place into the starch due to this process. Etherification using epoxide reagents is another significant way of starch modification. It lowers the hydrophilicity and inter- and intra-molecular H-bonding's degree in a starch strand by adding lipophilic (lipid soluble) alkyl groups (Tian et al. 2018).

6.6.1.2 Safety

Except in newborn formulae, modified starches can be the 43 constituent part of most meals according to good manufacturing practice (GMP). Maximum restrictions have been imposed on a few modified starches, such as distarch phosphate. In infant formulas and follow-up formulae, 5000 mg/kg distarch phosphate, acetylated distarch phosphate, acetylated distarch adipate, starch acetate, monostarch acetate while 50,000 mg/kg in supplemental meals for newborns and young children (Kumar et al. 2021).

6.6.1.3 Applications

Although native starches have been utilized as a gelling/thickener agent in numerous meals, their low surface activity restricts their utilization as food stabilizers. The employment of modified starches in the dairy, meat, canning, and spice industries has been documented owing to their viscofying and binding qualities, freezing-thawing stability, and smoothness (Singh et al. 2010).

6.6.2 Modified Cellulose (MC)

Cellulose—the utmost prevalent natural polymer on earth—is comprised of glucose repeating moieties allied together by β-1,4-linkages. It is the constituent part higher plant's cell walls, fungi, a few algae. Bacteria like *Acetobacter xylinum* also produce it. Sawdust, papaya peel, sago waste, cotton linters, water hyacinth, and sugarcane bagasse exemplify just a few agricultural products that advocate their use in cellulose's manufacturing. A few of the commercially available modified celluloses include carboxymethyl cellulose (CMC), methylcellulose (MC), hydroxypropyl methylcellulose, methyl ethyl cellulose (MEC), hydroxypropyl cellulose (HPC), microcrystalline cellulose (MCC), and ethyl cellulose (EC). Low molecular mass carbohydrates, proteins, waxes, and lignin are removed when the wood is pulped in alkali, allowing the cellulose to be recovered. It has to be modified for usage in food being water unsolvable in its natural form (Brigham et al. 2018).

6.6.2.1 Modification Procedures

Alkali-treated cellulose serves as the initial material for making modified celluloses. Carboxymethyl cellulose, a linear and anionic polysaccharide, is formed once alkali (NaOH/KOH)-treated cellulose reacts with monochloroacetic acid ($C_2H_3ClO_2$) in alkali and organic solvent's presence. This causes carboxymethyl group moieties in the cellulose strand/chain to partially replace hydroxyl groups, especially at the 2C atoms of glucose (Asl et al. 2017). The carboxymethyl moieties' number in CMC accounts for gum's solubility and degree of substitution. Wood/Cotton pulp cellulose is utilized to make HPMC and MC. Pulverized cellulose is mixed with dimethyl ether for producing a slurry. To make alkaline cellulose, sodium hydroxide is added gently to the slurry followed by mixing with methyl chloride (CH_3Cl) to make MC and propylene oxide (C_3H_6O) to make HPMC. During HPMC synthesis, the –OH groups in anhydrous glucose moieties are replaced with 2-hydroxypropyl ($C_7H_{12}O_3$) groups. For MC synthesis, the –OH groups at 2, 3, and 6 carbon positions in anhydrous glucose moieties are substituted with methyl (CH_3) groups. MCC is the fourth most common modified cellulose, and it is synthesized from cellulose that has been hydrolyzed under synchronized conditions. This hydrolysis could be accomplished through enzymatic reaction, weak acid treatment, extrusion cooking, or steam explosion. All these treatments abolish cellulose's amorphous sections, left behind crystalline domains only. Attributable to the tiny particles production and the shorter reaction time, acid hydrolysis is preferred as comparable to other treatments (Chauhan et al. 2010). The properties exhibited by the modified celluloses are portrayed in Table 6.3.

6.6.2.2 Safety

The European Food Safety Authority (EFSA), during re-evaluating cellulose(s) as a food additive(s), determined that cellulose and its derivative's severe toxicity was minimal and food that uses modified cellulose up to 10% concentration had no adverse effects. For cellulose derivatives, the NOAEL (no-observed-adverse-effect-level) has been calculated to be 9000 mg/kg BW/day. This panel also decided that cellulose and its derivatives pose no safety risk at the reported levels of various applications. The panel used a cumulative exposure dosage of 660–900 mg/kg BW/day for modified celluloses (Kumar et al. 2021).

6.6.2.3 Applications

The most recurrently employed modified celluloses as food stabilizers include MC, HPMC, MCC, and CMC. The purified form of CMC powder is utilized to prepare ice cream, pastries, and drinks. MC is utilized to manufacture gluten-free breads besides improving their texture. It is also employed in mayonnaise to keep emulsion stability, regulate viscosity, and stabilize foam in cold beverages. HPMC is reported for its frequent utilization in the food sector for reducing oil absorption in fried foods. MCC is utilized as an emulsifier, thickener, texturizer, bulking, and anticaking agent. It has also been utilized to manufacture fat-free ice cream, pastry items, and slimming products (Deshmukh et al. 2017).

6.7 Food Stabilizers from Genetically Engineered Microbes

Microorganisms that are genetically edited utilizing in vitro biotechnological techniques to execute a particular function are acknowledged as genetically engineered microorganisms/microbes (GEMs) which might be either bacteria or fungi. The usage of GEMs in manufacturing food ingredients and food processing aids has become pervasive in the latest few decades (Hanlona and Sewalt 2021). Examples include vitamins, functional proteins (texturants), amino acids, nutritional proteins, sweeteners, flavors, and oligosaccharides (Adrio and Demain 2010).

Among GEMs, *Actinobacillus succinogenes, Basfia succiniciproducens, Mannheimia succiniciproducens*, and *Escherichia coli* are being extensively utilized GEMs for producing various food stabilizers, primarily carboxylic and amino acids whose production is allied with carbon metabolism in them. According to literature sources carboxylic acids especially aliphatic are typically utilized as acidity regulators or preservatives. Many of the carboxylic acids, including lactic, succinic, citric, malic, and fumaric acids that are produced via tricarboxylic acid cycle and glycolysis, have been recorded in the E-numbers catalogue. Lactic acid (E270) is produced by numerous microbes like bacteria (Lactic acid bacteria—*Lactococcus* spp. and

Lactobacillus spp.; *Corynebacterium glutamicum*, *Escherichia coli*, *Bacillus subtilis*) microalgae, and yeasts (Abdel-Rahman et al. 2013) whereas malic acid (E296)—another intermediate of TCA cycle—is typically generated by the genus Aspergillus, viz. *A. oryzae*, *A. flavus*, and *A. niger*. The TCA cycle's reductive phase has also been tweaked to enhance malic acid's synthesis from pyruvate synthesized from glycolysis. When the genes encoding for C4-dicarboxylate transporter, pyruvate carboxylase, and malate dehydrogenase were overexpressed in a genetically engineered *A. oryzae* NRRL 3488 strain, synthesis of malic acid (154 g/L) and glucose (1.02 g/g) was reported which was 69% of the highest yield on theoretical basis (Brown et al. 2013).

For malic acid synthesis, analogous metabolic engineering procedures were used in *Ustilago trichophora, E. coli*, and *S. cerevisiae*. The L-glutamic acid (E620), L-cysteine (E640), and glycine (E640) have all been licensed by the EU as flavor enhancers (E920). The main chemical accountable for the "umami" flavor of meals is the monosodium salt form of L-glutamic acid (E621). L-Cysteine is mainly utilized as a flour treatment agent for improving baked item's functionality. All of the three above-mentioned amino acids may theoretically be the derivatives of either hydrolyzed protein or GEMs. Glycine is an amino acid whose synthesis is easier by chemical techniques; consequently neither methodology is employed for industrial manufacturing. Because glycine lacks stereogenic centers, it does not necessitate the costly procedure of separating the L-forms from a racemic mix of chemically generated amino acids (Zeng et al. 2016). The mechanism of glutamate synthesis was thoroughly explored in the subsequent decades accompanied by a systematic engineering of *C. glutamicum* to enhance product titers values (Hirasawa and Wachi 2016). A strain exhibiting the ability of producing 100 g/L glutamic acid (L-form) and glucose (yield—0.6 g per g) was developed through a blend of strain engineering techniques and process parameter's optimization. Several publications and review papers (Kimura 2003; Sano 2009; Eggeling and Bott 2005) are reported to outline metabolic engineering headways to increase L-glutamic acid formation in *C. glutamicum*.

Protein hydrolysis from animal sources like poultry feathers is presently the most prevalent source of L-cysteine synthesis. To make engineered L-cysteine, researchers used the natural biosynthetic process starting with L-serine and acetyl-CoA in *C. glutamicum* and *E. coli* (Wada and Takagi 2006). Metabolic engineering aimed to mutate the cysE gene in order to develop enzymes that were little feedback sensitive (Kai et al. 2006). Functionally introducing heterologous and feedback-insensitive CysEiso enzymes into *Escherichia coli* from *Arabidopsis thaliana* produced 1.7 g/L titer of L-cysteine (Takagi et al. 1999). During a recent investigation, omission of yciW gene that codes for an oxidoreductase enzyme apparently tangled in metabolizing cysteine (L-form) in *E. coli* ensued an upsurge in L-cysteine production, despite the set titer value (1.7 g/L) in 1999 stood surpassed (Takagi and Ohtsu 2016).

6.8 Concluding Remarks

Microorganisms, including GEMs, use an array of enzymes and metabolic processes to make microbial polysaccharides like dextran, gellan gum, pollulan, xanthan gum, scleroglucan along with other natural food stabilizers that are healthier than synthetic ones. This is endorsed by the point that synthetic food stabilizers have a negative impact on consumer health owing to the employment of harsh chemicals in their manufacturing besides the development of hazardous waste. Microbial stabilizers, conversely, have a positive influence as they are often synthesized through fermentative methods, making them a practical and environmentally acceptable option. Furthermore, techniques employing genetic engineering tools have been acknowledged to be particularly successful for large-scale production by directing bacteria to synthesize desired molecules. Consequently, microbial food stabilizers have risen in demand in the international market. However, the primary disadvantage allied with the production of microbial polysaccharide is its high cost. Researchers must employ an assorted and comprehensive research approach that integrates approaches from both industrial biotechnology and systems biology if microbial food stabilizers are to play their rightful role in the food business. In a nutshell, growth of food stabilizers in the international market is reliant on a multidisciplinary scientific approach and alternative production strategies for large-scale manufacturing of microbial food stabilizers to bear out food trade's necessities.

References

Abdel-Rahman MA, Tashiro Y, Sonomoto K (2013) Recent advances in lactic acid production by microbial fermentation processes. Biotechnol Adv 31:877–902. https://doi.org/10.1016/j.biotechadv.2013.04.002

Adrio JL, Demain AL (2010) Recombinant organisms for production of industrial products. Bioeng Bugs 1:116–131. https://doi.org/10.4161/bbug.1.2.10484

Ahmad NH, Shuhaimi M, Yaakob B (2015) Microbial polysaccharides and their modification approaches: a review. Int J Food Prop 18:332–347

Aman A, Siddiqui NN, Qader SAU (2012) Characterization and potential applications of high molecular weight dextran produced by *Leuconostoc mesenteroides* AA1. Carbohydr Polym 87: 910–915

Andhare P, Chauhan K, Dave M et al (2014) Microbial exopolysaccharides: advances in applications and future prospects. In: Tewari R (ed) Biotechnology, 3: microbial biotechnology. Studium Press, New Delhi, pp 1–25

Asl SA, Mousavi M, Labbafi M (2017) Synthesis and characterization of carboxymethyl cellulose from sugarcane bagasse. J Food Process Technol 8:1–6. https://doi.org/10.4172/2157-7110.1000687

Bajaj IB, Survase SA, Saudagar PS et al (2007) Gellan gum: fermentative production, downstream processing and applications. Food Technol Biotechnol 45:341–354

Barak S, Mudgil D (2014) Locust bean gum: processing, properties and food applications—a review. Int J Biol Macromol 66:74–80. https://doi.org/10.1016/j.ijbiomac.2014.02.017

Bhavani AL, Nisha J (2010) Dextran—the polysaccharide with versatile uses. Int J Pharma Bio Sci 1:569–573

Brigham JS, Adney WS, Himmel ME (2018) Hemicellulases: diversity and applications. In: Wayman CE (ed) Handbook on bioethanology. Routledge, Milton Park, pp 119–141

Brown SH, Bashkirova L, Berka R, Chandler T, Doty T, McCall K et al (2013) Metabolic engineering of *Aspergillus oryzae* NRRL 3488 for increased production of L-malic acid. Appl Microbiol Biotechnol 97:8903–8912. https://doi.org/10.1007/s00253-013-5132-2

Celik GY, Aslim B, Beyatli Y (2008) Characterization and production of the exopolysaccharide (EPS) from *Pseudomonas aeruginosa* G1 and *Pseudomonas putida* G12 strains. Carbohydr Polym 73:178–182

Chauhan YP, Khedkar SV, Bhagat SL et al (2010) A comparative study of acid hydrolysis of cellulosic waste (waste of hosiery industry) for manufacturing microcrystalline cellulose. Int J Chem Sci 8:2227–2235

Chiu C, Solarek D (2009) Modification of starches. In: BeMiller JN, Whistler RL (eds) Starch: chemistry and technology. Elsevier, Burlington, pp 629–655

Choi S, Choi W, Kim S et al (2014) Purification and biocompatibility of fermented hyaluronic acid for its applications to biomaterials. Biomater Res 18:6

Coviello T, Matricardi P, Marianecci C et al (2007) Polysaccharide hydrogels for modified release formulations. J Control Release 119:5–24

Deshmukh K, Ahamed MB, Deshmukh RR et al (2017) Newly developed biodegradable polymer nanocomposites of cellulose acetate and Al_2O_3 nanoparticles with enhanced dielectric performance for embedded passive applications. J Mater Sci Mater Electron 28(1):973–986. https://doi.org/10.1007/s10854-016-5616-9

Díaz-Montes E (2021) Dextran: sources, structures and properties. Polysaccharides 2:554–565

Echave MC, Hernáez-Moya R, Iturriaga L et al (2019) Recent advances in gelatin-based therapeutics. Expert Opin Biol Ther 19:773–779. https://doi.org/10.1080/14712598.2019.1610383

Eggeling L, Bott M (eds) (2005) Handbook of Corynebacterium glutamicum, 1st edn. CRC Press. https://doi.org/10.1201/9781420039696

Egharevba HO (2019) Chemical properties of starch and its application in the food industry. In: Emeje M (ed) Chemical properties of starch. IntechOpen, Shanghai, pp 63–88

Elnour AA, Mirghani ME, Kabbashi NA et al (2018) Study of antioxidant and anti-inflammatory crude methanol extract and fractions of acacia seyal gum. Am J Pharmacol Pharmacother 5(1):3. https://doi.org/10.21767/2393-8862.100012

Falconer DJ, Mukerjea R, Robyt JF (2011) Biosynthesis of dextrans with different molecular weights by selecting the concentration of *Leuconostoc mesenteroides* B-512FMC dextransucrase, the sucrose concentration and the temperature. Carbohydr Polym 364:280–284

FAO, WHO (2020) Compendium of food additive specifications. Joint FAO/WHO Expert Committee on Food Additives (JECFA), 87th meeting June 2019. FAO JECFA monographs 23. Rome. https://doi.org/10.4060/ca7513en

Ferreira ARV, Alves VD, Coelhoso IM (2016) Polysaccharide-based membranes in food packaging applications. Membranes 6:1–17

Freitas F, Torres CA, Araújo D et al (2021) Advanced microbial polysaccharides. Biopolym Biomed Biotechnol Appl 1:19–62

FSSAI Act (2006) Rules (2011) Akalank's food safety and standards act, rules and regulation. Akalank Publication, New Delhi, p 293

Ghanbarzadeh B, Almasi H (2013) Biodegradable polymers. Biodegrad Life Sci 1:141–185

Hanlona P, Sewalt V (2021) GEMs: genetically engineered microorganisms and the regulatory oversight of their uses in modern food production. Crit Rev Food Sci Nutr 61:959–970

Heinze T, Liebert T, Heublein B et al (2006) Functional polymers based on dextran. In: Klemm D (ed) Polysaccharides II. Springer, Berlin, pp 199–291

Hirasawa T, Wachi M (2016) Glutamate fermentation-2: mechanism of L-glutamate overproduction in *Corynebacterium glutamicum*. In: Yokota A, Ikeda M (eds) Amino acid fermentation. Springer, Berlin, pp 57–72

Huang HC, Chen LC, Lin SB (2010) *In situ* modification of bacterial cellulose network structure by adding interfering substances during fermentation. Bioresour Technol 101:6084–6091

Imeson A (2010) Food stabilizers, thickeners and gelling agents. Wiley-Blackwell, Oxford

Jain A, Gupta Y, Jain SK (2007) Perspectives of biodegradable natural polysaccharides for site-specific drug delivery to the colon. J Pharm Pharmaceut Sci 10:86–128

Jeewanthi RK, Lee NK, Paik HD (2015) Improved functional characteristics of whey protein hydrolysates in food industry. Korean J Food Sci Anim Res 353:350–359. https://doi.org/10.5851/kosfa.2015.35.3.350

Jia F, Liu HJ, Zhang GG (2016) Preparation of carboxymethyl cellulose from corncob. Procedia Environ Sci 31(1):98–102. https://doi.org/10.1016/j.proenv.2016.02.013

Jin Y, Zhang H, Yin Y et al (2006) Conformation of curdlan as observed by tapping mode atomic force microscopy. Colloid Polym Sci 284:1371–1377

Jindal N, Khattar JS (2018) Microbial polysaccharides in food industry. In: Biopolymers for food design. Academic, New York, pp 95–123

Kai H, Mori T, Tokuda K, Takayama N, Tahara N, Takemiya K, Kudo H, Sugi Y, Fukui D, Yasukawa H, Kuwahara F, Imaizumi T (2006) Pressure overload-induced transient oxidative stress mediates perivascular inflammation and cardiac fibrosis through angiotensin II. Hypertens Res 29:711–718

Kimura T, Goto T, Shintani H et al (2003) Magnetic control of ferroelectric polarization. Nature 426:55–58. https://doi.org/10.1038/nature02018

Kırtel O, Avşar G, Erkorkmaz BA et al (2017) Microbial polysaccharides as food ingredients. In: Microbial production of food ingredients and additives. Academic, New York, pp 347–383

Korma SA, Niazi S, Ammar AF et al (2016) Chemically modified starch and utilization in food stuffs. Int J Nutr Food Sci 5:264–272. https://doi.org/10.11648/j.ijnfs.20160504.15

Kothari D, Das D, Patel S, Goyal A (2014) Dextran and food application. Springer. https://doi.org/10.1007/978-3-319-03751-6_66-1

Krempel M, Griffin K, Khouryieh H (2019) Hydrocolloids as emulsifiers and stabilizers in beverage preservation. In: Grumezescu A, Holban AM (eds) Preservatives and preservation approaches in beverages. Academic, Cambridge, pp 427–465

Kumar A, Kumari P, Gupta K et al (2021) Recent advances in extraction, techno-functional properties, food and therapeutic applications as well as safety aspects of natural and modified stabilizers. Food Rev Int 1:1–44

Lee IY (2002) Curdlan. In: Vandamme EJ, De Baets S, Steinbuchel A (eds) Polysaccharides I: polysaccharides from prokaryotes, Biopolymers, vol 5. Wiley-VCH, Weinheim, pp 135–158

Li Y, Li C, Gu Z (2019) Digestion properties of corn starch modified by α-D-glucan branching enzyme and cyclodextrin glycosyltransferase. Food Hydrocoll 89:534–541

Liang TW, Wang SL (2015) Recent advances in exopolysaccharides from *Paenibacillus* spp.: production, isolation, structure and bioactivities. Mar Drugs 13:1847–1863

Lopes BM, Lessa VL, Silva BM et al (2015) Xanthan gum: properties, production conditions, quality and economic perspective. J Food Nutr Res 54:185–194

Luckett CR, Wang YJ (2012) Application of enzyme-treated corn starches in breakfast cereal coating. J Food Sci 77:C901–C906. https://doi.org/10.1111/j.1750-3841.2012.02794.x

Maddern EC (2008) Porcine and bovine surgical products, Jewish, Muslim and Hindu perspectives. Arch Surg 143:366–370

Majumder A, Goyal A (2009) Rheological and gelling properties of novel glucan from *leuconostoc dextranicum* NRRL B-1146. Int Food Res J 42:525–528

Mariod AA, Fadul H (2013) Gelatin, source, extraction and industrial applications. Acta Sci Pol 12: 135–147

McIntosh M, Stone BA, Stanisich VA (2005) Curdlan and other bacterial (1→3)-β-D-glucans. Appl Microbiol Biotechnol 68:163–173

Mudgil D, Barak S, Khatkar BS (2015) Effect of partially hydrolyzed guar gum on pasting, thermo-mechanical and rheological properties of wheat dough. Int J Biol Macromol 93:131–135. https://doi.org/10.1016/j.ijbiomac.2016.08.064

Naessens M, Cerdobbel A, Soetaert W et al (2005) *Leuconostoc* dextran, sucrase and dextran: production, properties and applications. J Chem Technol Biotechnol 80:845–860. https://doi.org/10.1002/jctb.1322

Narayanaswamy R, Kanagesan S, Pandurangan AK et al (2016) Basics to different imaging techniques, different nanobiomaterials for image enhancement. In: Grumezescu AM (ed) Nanobiomaterials in medical imaging. William Andrew Publishing, Norwich, pp 101–129

Nasatto PL, Pignon F, Silveira JL (2015) Methylcellulose, a cellulose derivative with original physical properties and extended applications. Polymers 7:777–803. https://doi.org/10.3390/polym7050777

Nasrollahzadeh M, Sajjadi M, Sajadi SM et al (2019) Green nanotechnology. In: Nasrollahzadeh M, Sajadi SM, Sajjadi M, Issaabadi Z, Atarod M (eds) Interface science and technology. Elsevier, Amsterdam, pp 145–198. https://doi.org/10.1016/B978-0-12-813586-0.00005-5

Nasrollahzadeh M, Sajjadi M, Nezafat Z et al (2021) Polysaccharide biopolymer chemistry. In: Nasrollahzadeh M (ed) Biopolymer based metal nanoparticle chemistry for sustainable applications. Elsevier, Amsterdam, pp 45–105. https://doi.org/10.1016/B978-0-12-822108-2.00019-3

Navard P (2013) The European polysaccharide network of excellence (EPNOE). Springer-Verlag GmbH, Wien

Nishinari K, Zhang H, Funami T (2021) Curdlan. In: Handbook of hydrocolloids. Woodhead Publishing, Sawston, pp 887–921

Oliveira JT, Martins L, Picciochi R et al (2010) Gellan gum: a new biomaterial for cartilage tissue engineering applications. J Biomed Mater Res A 93:852–863

Osmałek T, Froelich A, Tasarek S (2014) Application of gellan gum in pharmacy and medicine. Int J Pharm 466:328–340. https://doi.org/10.1016/j.ijpharm.2014.03.038

Philp K (2018) Polysaccharide ingredients. In: Reference module in food science. Elsevier, Amsterdam, pp 1–23. https://doi.org/10.1016/b978-0-08-100596-5.22367-6

Qin Y, Zhang H, Dai Y et al (2019) Effect of alkali treatment on structure and properties of high amylose corn starch film. Materials (Basel) 12:1705. https://doi.org/10.3390/ma12101705

Quinn FX, Hatakeyama T, Yoshida H et al (1993) The conformational properties of gellan gum hydrogels. Polym Gels Netw 1:93–114

Roy JC, Salaün F, Giraud S (2017) Solubility of chitin: solvents, solution behaviors and their related mechanisms. In: Zhenbo X (ed) Solubility of polysaccharides. IntechOpen, Shanghai, pp 109–128. https://doi.org/10.5772/intechopen.71385

Saha D, Bhattacharya S (2010) Hydrocolloids as thickening and gelling agents in food: a critical review. J Food Sci Technol 47:587–597

Sahu P, Pisalkar PS, Patel S et al (2019) Physico-chemical and rheological properties of Karaya gum (Sterculia Urens Roxb.). Int J Curr Microbiol App Sci 8:672–681. https://doi.org/10.20546/ijcmas.2019.804.072

Sano C (2009) History of glutamate production. Am J Clin Nutr 90:728S–732S. https://doi.org/10.3945/ajcn.2009.27462F

Sarbini SR, Kolida S, Naeye T et al (2013) The prebiotic effect of α-1,2 branched, low molecular weight dextran in the batch and continuous faecal fermentation system. J Funct Foods 5:1938–1946

Schmid J, Meyer V, Sieber V (2011) Scleroglucan: biosynthesis, production and application of a versatile hydrocolloid. Appl Microbiol Biotechnol 91:937–947

Singh V, Ali SZ (2008) Properties of starches modified by different acids. Int J Food Prop 11:495–507. https://doi.org/10.1080/10942910802083774

Singh S, Singh N, Isono N, Noda T (2010) Relationship of granule size distribution and amylopectin structure with pasting, thermal, and retrogradation properties in wheat starch. J Agric Food Chem 58:1180–1188. https://doi.org/10.1021/jf902753f

Sousa AMM, Borges J, Silva AF et al (2013) Influence of the extraction process on the rheological and structural properties of agars. Carbohydr Polym 96:163–171

Spicer EJF, Goldenthal EI, Ikeda T (1999) A toxicological assessment of curdlan. Food Chem Toxicol 37:455–479

Stanley NF (2006) Agars. In: Stephen AM, Philips GO, Williams PA (eds) Food polysaccharides and their applications, 2nd edn. CRC Press, Boca Raton, pp 217–230

Stephen AM, Phillips GO, Williams PA (2006) Food polysaccharides and their applications, 2nd edn. CRC Press, Boca Raton

Survase SA, Saudagar PS, Bajaj IB et al (2007) Scleroglucan: fermentative production, downstream processing and applications. Food Technol Biotechnol 45:107–118

Szekalska M, Puciłowska A, Szymańska E (2016) Current use and future perspectives in pharmaceutical and biomedical applications. Int J Polym Sci 2016(8):1–17. https://doi.org/10.1155/2016/7697031

Takagi H, Ohtsu I (2016) L-cysteine metabolism and fermentation in microorganisms. In: Yokota A, Ikeda M (eds) Amino acid fermentation. Springer, Berlin, pp 129–151

Takagi H, Awano N, Kobayashi SI et al (1999) Overproduction of L-cysteine and L-cystine by expression of genes for feedback inhibition-insensitive serine acetyltransferase from *Arabidopsis thaliana* in *Escherichia coli*. FEMS Microbiol Lett 179:453–459. https://doi.org/10.1016/S0378-1097(99)00429-2

Tarande MG, Manriquez-Hernandez J (2016) Carrageenan properties and applications: a review. In: Pereira L (ed) Carrageenans: sources and extraction methods, molecular structure, bioactive properties and health effects. Nova Science, Hauppauge, pp 17–49

Tian S, Chen Y, Chen Z et al (2018) Preparation and characteristics of starch esters and its effects on dough physicochemical properties. J Food Qual 2018(12):1–7. https://doi.org/10.1155/2018/1395978

Utami R, Kawiji, Nurhartadi E et al (2014) The effect of cassava starch-based edible coating enriched with *Kaempferia rotunda* and *Curcuma xanthorrhiza* essential oil on refrigerated patin fillets quality. Int Food Res J 21:413–419

Vanitha T, Khan M (2019) Role of pectin in food processing and food packaging. In: Martin M (ed) Pectins—extraction, purification, characterization and applications. IntechOpen, Shanghai, pp 85–104. https://doi.org/10.5772/intechopen.83677

Venugopal V (2011) Polysaccharides: their characteristics and marine sources. In: Marine polysaccharides food applications. CRC Press, Boca Raton, pp 3–27

Wada M, Takagi H (2006) Metabolic pathways and biotechnological production of L-cysteine. Appl Microbiol Biotechnol 73:48–54. https://doi.org/10.1007/s00253-006-0587-z

Wang A, Lewus R, Rathore AS (2006) Comparison of different options for harvest of a therapeutic protein product from high cell density yeast fermentation broth. Biotechnol Bioeng 94:91–104

Wolter A, Hager A, Zannini E et al (2014) Influence of dextran-producing *Weissella cibaria* on baking properties and sensory profile of gluten-free and wheat breads. Int J Food Microbiol 172: 83–91

Wu Y, Ding W, Jia L et al (2015) The rheological properties of Tara gum (Caesalpinia spinosa). Food Chem 168:366–371. https://doi.org/10.1016/j.foodchem.2014.07.083

Wustenberg T (2014) General overview of food hydrocolloids. In: Cellulose and cellulose derivatives in the food industry: fundamentals and applications. Wiley-VCH, Weinheim, pp 1–68

Zarour K, Llamas MG, Prieto A et al (2017) Rheology and bioactivity of high molecular weight dextrans synthesised by lactic acid bacteria. Carbohydr Polym 174:646–657

Zeng Y, Li Z, Demopoulos GP (2016) Process for glycine production by antisolvent crystallization using its phase equilibria in the ethylene glycol–NH4Cl–water system. Ind Eng Chem Res 55: 2426–2437. https://doi.org/10.1021/acs.iecr.5b04144

Zhang Y, Kong H, Fang Y et al (2013) Schizophyllan: a review on its structure, properties, bioactivities and recent developments. Bioact Carbohydr Diet Fibre 1:53–71

Chapter 7
Biosurfactants: Promising Biomolecules in the Food Industry

Beatriz Galdino Ribeiro, Jenyffer Medeiros Campos Guerra, and Leonie Asfora Sarubbo

Abstract Biosurfactants are molecules derived from microorganisms that have potential for application in many sectors of the industry. This is due to several advantages, such as the action of detergency, emulsification, dispersions foaming or solubilization of phases, high biodegradability, and production from different renewable sources. Several microorganisms can produce biosurfactants as by-products of their metabolism, including bacteria and yeasts. In the food sector, these biomolecules have aroused increasing interest due to antimicrobial and antiadhesive properties, important to prevent and/or minimize contamination and increase the shelf life of foods, as well as antioxidant properties, which add even more value to formulations of microbial surfactants, in addition to contributing to the overall quality control. In addition, the possibility of production by microorganisms with GRAS status (Generally Regarded as Safe) guarantees the safety of their use due to their low toxicity and proven pathogenicity. In food processing, the ability of biosurfactants to act as emulsifiers and emulsion stabilizers allows them to be used in different concentrations, as additives and even versatile ingredients. Among the potential food formulations, ice cream, salad dressings, breads, cookies and muffins stand out, which, when incorporated with microbial surfactants, showed improvements in physical and physical-chemical properties and in the analysis of the texture profile. Thus, this chapter discusses the biosurfactants application prospects in food formulations, highlighting improvements in emulsification, stabilization and textural and

B. G. Ribeiro
Northeast Biotechnology Network (RENORBIO), Federal Rural University of Pernambuco, Recife, Pernambuco, Brazil

J. M. Campos Guerra
Department of Chemical Engineering, Federal University of Pernambuco, Recife, Pernambuco, Brazil

L. A. Sarubbo (✉)
Icam Tech School, Catholic University of Pernambuco, Recife, Pernambuco, Brazil

Advanced Institute of Technology and Innovation (IATI), Recife, Pernambuco, Brazil
e-mail: leonie.sarubbo@unicap.br

A. K. Nadda, G. Goel (eds.), *Microbes for Natural Food Additives*, Microorganisms for Sustainability 38, https://doi.org/10.1007/978-981-19-5711-6_7

sensory aspects of different types of food, as well as the technological and economic challenges associated with the use of microbial surfactants in this promising sector.

Keywords Biodegradability · Emulsification · Foods formulation · Microorganisms · Surfactants

7.1 Introduction

Biosurfactants are known as ecologically friendly molecules, as they are biodegradable, biocompatible, and show low toxicity, in addition to standing out for having a wide range of applications. Like synthetic surfactants, biosurfactants are surface-active agents with amphipathic properties, with a polar and a nonpolar region, which can be used in hydrophobic and/or hydrophilic systems, where they act in processes of detergency and dispersion, wetting, foaming, and emulsification (Farias et al. 2021; Ohadi et al. 2019).

Increased research on biosurfactants confirms their high efficiency in the processes in which they are used, mainly in the hydrocarbon bioremediation and microbial enhanced oil recovery (MEOR) sectors (Sales da Silva et al. 2020). In addition, their unique characteristics allow them to be applied in a wide range of sectors, such as cleaning products and detergents (54%), textiles (13%), chemical processes (10%), leather and paper (10%), cosmetics and pharmaceuticals (3%), food (3%), and agriculture (2%), with the function of replacing, at least partially, chemical surfactants, contributing positively to industry and responsibly for the environment (Farias et al. 2021; Shakeri et al. 2021).

The production of biosurfactants can be carried out by the metabolism of many microorganisms from hydrophobic and hydrophilic substrates. Among the producers' microorganisms, bacteria of the genus *Bacillus* and *Pseudomonas* and yeasts of the genus *Candida*, *Starmerella*, and *Saccharomyces* stand out. The composition and structure of a biosurfactant is influenced both by the substrates present in the medium and by the cultivation conditions. These parameters are objects of extensive investigation to direct the production of these biomolecules according to specific applications. Glycolipids and lipopeptides are the most known types of biosurfactants due to their versatile physicochemical properties and differentiated biological activities, with excellent surface-active action, even under extreme conditions (Mnif et al. 2017).

Biosurfactants also show antimicrobial, antiadhesive, and antioxidant activities, which make them interesting for the food industry by promoting increased food safety and, consequently, general quality of the products (Silva and Sarubbo 2021). In addition to contributing to reduce contamination and microbial adhesion and increase the shelf life of final formulations, by preventing the occurrence of oxidation reactions, biosurfactants also influence the efficient removal of heavy metals from the surface of vegetables by acting as washing agents (Liu et al. 2019).

The benefits of applying these biomolecules especially in the food sector are due to their broad emulsification spectrum. Biosurfactants can effectively emulsify and stabilize vegetable oils such as castor, mustard, coconut, sesame, and sunflower, as well as improve some important characteristics such as viscosity, volume, texture and even energy value of ice cream, salad dressings, breads, cakes, and cookies. However, the difficulties in applying biosurfactants in the food industry are still real, whether due to the high cost of production, separation, and purification, or the low efficiency in some types of application. Thus, it is necessary that studies are focused on both synthesis technology and large-scale application so that biosurfactants can establish themselves in this sector (Ribeiro et al. 2020a).

In this chapter, the properties, classification, and production of biosurfactants are discussed, focusing on the most recent approaches to the application of these biomolecules in the food industry, considering the technological and economic challenges of their use in this expanding sector.

7.2 Biosurfactants

Biosurfactants are tensioactive agents with amphiphilic nature, which have a hydrophilic and a hydrophobic portion in their structure. The main difference with synthetic surfactants is due to the possibility of production from microorganisms. Their application is quite wide, due to the ability to reduce the surface and interfacial tension of various media through the formation of micelles, which accumulate at the interface between liquids of different polarities, such as water and oil (Karlapudi et al. 2018; Ohadi et al. 2019) (Fig. 7.1). Since biosurfactants originate from the secondary metabolism of microorganisms and show performance equivalent to synthetic surfactants, biosurfactants are considered more advantageous due to the growing search for more sustainable and less aggressive surfactants to the environment (Najmi et al. 2018; Santos et al. 2018).

Biosurfactants play a significant role in the formation of direct and reverse microemulsions, clear and stable liquid mixtures composed of water and oil, separated by monolayers or aggregates of surfactants, being formed when a liquid phase is dispersed as droplets in another liquid phase (Perfumo et al. 2017; Rocha e Silva

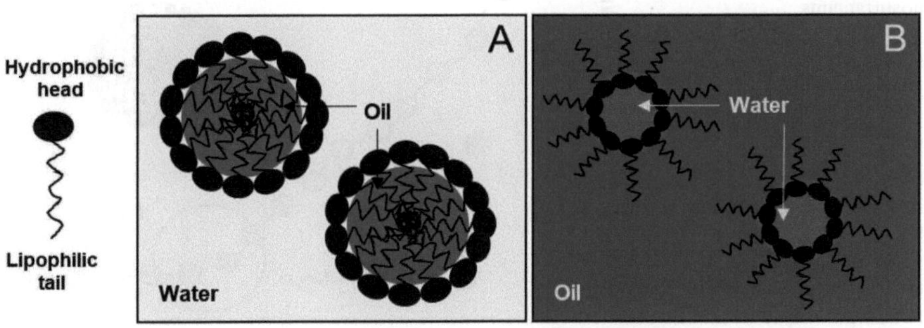

Fig. 7.1 Biosurfactant molecules forming micelles in oil-in-water (**a**) and water-in-oil (**b**) systems

et al. 2019). This is due to the formation of micelles, which are amphipathic molecules aggregated to hydrophilic and hydrophobic portions that increase the solubility of organic compounds in the aqueous phase, favoring the application of biosurfactants in several areas. They can also be applied under specific conditions, since, in addition to low toxicity and high biodegradability, they are stable in the face of significant variations in pH, temperature, and salt concentration (Rocha e Silva et al. 2018; Anic et al. 2018).

7.3 Properties of Biosurfactants

Among the properties of a biosurfactant, detergency, emulsification, dispersion, foaming, or solubilization of different phases stands out, being directly or indirectly related to the ability to reduce the surface tension of a system, the reduction being inversely proportional to its concentration in the medium (Maddikeri et al. 2015; Bakhshi et al. 2017). Surface tension, which measures the effectiveness of the biosurfactant along with the interfacial tension of a system, decreases as micelles are formed, and its concentration is related to the Critical Micelle Concentration (CMC), which corresponds to the minimum concentration of surfactant necessary for that the surface and interfacial tension is reduced to the maximum, being an efficiency parameter of a biosurfactant (Fig. 7.2) (Araújo et al. 2020; Gouda et al. 2020).

The effectiveness and efficiency of a biosurfactant can also be influenced by molecular weight. Biosurfactants that have low molecular weight are more efficient in reducing surface and interfacial tension, contributing to solubilization of immiscible compounds. Structures that present high molecular weight, on the other hand, are more effective and efficient in stabilizing oil-water emulsions, quite common in

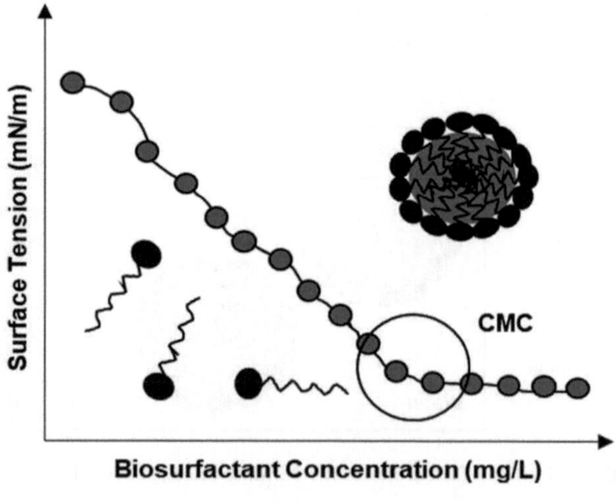

Fig. 7.2 Relationship between surface tension and the critical micelle concentration (CMC) of biosurfactants

foods, reducing repulsive forces and, thus, allowing easier mixing and interaction between the two phases (Fari and Saeed 2016; Solomon et al. 2017).

In addition to playing an important role in the solubilization of hydrophobic compounds (Zhong et al. 2016; Liu et al. 2017), microbial surfactants can act in the binding of heavy metals (Chen et al. 2017; Arab and Mulligan 2018) and in the production of antimicrobial compounds, antibiofilms, antioxidants, and antibiotics (Kaczorek et al. 2018; Lima et al. 2019). Biosurfactants can also have other interesting properties, such as hemolytic, antiviral, anticancer, and immunomodulatory action, being able to be used in biomedicine and therapy (Mnif and Ghribi 2016; Yalaoui-Guellal et al. 2020; Marangon et al. 2020). These properties make biosurfactants a safe alternative to combat various foodborne pathogens (Sharma and Saharan 2016; Ekpenyong et al. 2017).

7.4 Classification of Biosurfactants

The main classification criterion for microbial surfactants is the type of chemical structure presented. The most common biosurfactants are glycolipids, lipoproteins, lipopolysaccharides, complex biopolymers, phospholipids, and lipopeptides (Souza et al. 2017). However, other parameters are also considered, such as molecular weight—high (lipoproteins, lipopolysaccharides, and complex biopolymers) or low (glycolipids, lipopeptides, and phospholipids), ionic charges (anionic, cationic, nonionic, and neutral), and secretion type (intracellular, extracellular, and adherent to microbial cells). As for production and potential applications in food, glycolipids and lipopeptides are the most studied (Mnif et al. 2017).

Glycolipids are composed of fatty acids combined with carbohydrates through glycosidic bonds. They are subdivided into several types, according to the carbohydrate present in the structure. The best-known glycolipids are sophorolipids, trehalolipids, mannosylerythritol (MEL) lipids, and rhamnolipids (Williams and Trindade 2017). This subdivision occurs according to the nature of the carbohydrate and lipid fraction in the molecule, as well as according to the producing microorganism, with rhamnolipids commonly coming from *Pseudomonas* sp. (Shreve and Makula 2019; Varjani and Upasani 2019) and *Saccharomyces* (Bahia et al. 2018), trehalolipids produced by *Rhodococcus* and other bacterial species (Kuyukina and Ivshina 2019), mannosylerythritol lipids by *Pseudozyma* sp. (Fukuoka et al. 2016) and sophorolipids and carbohydrate-protein-lipid complexes produced mainly by yeasts of the *Candida* genus (Gaur et al. 2019; Campos et al. 2019).

The structure of sophorolipids has a sophorose molecule linked to a long-chain hydroxyl fatty acid through a glycosidic bond (Ozdener et al. 2019). Trehalolipids are composed of mycolic acid (long-chain β-hydroxycarboxylic acid) linked to trehalose disaccharide and are also produced by bacterial species *Mycobacterium*, *Nocardia*, and *Corynebacterium*. Mannosylerythritol (MEL) lipids are made up of mannose sugar, linked to the fatty acid chain, which may differ in relation to the degree of saturation, length of the hydrophobic chain, and/or acetylation in the C4 and C6 positions of the sugar (Bages-Estopa et al. 2018). Rhamnolipids, on the other hand, consist of rhamnose units linked to 3-hydroxyl fatty acid units via a β-glycosidic link. Rhamnose units are linked together by *O*-glycosidic bonds, whereas fatty acids are linked by ester-type bonds (Irorere et al. 2017).

The structure of lipopeptides differs from that of glycolipids by replacing carbohydrate molecules with peptide molecules linked to the lipid tail, which can be linear or cyclic. Lipopeptides have some interesting properties, such as biological activity, low toxic potential, and efficacy even at low concentrations, proving to be suitable for medical, cosmetic, and food applications. One of the examples is Surfactin, the main class of lipopeptides, with anionic charge, composed of seven D and L amino acid residues and a fatty acid residue containing 13–15 carbon atoms, whose studies show a significant surface-active activity, with the reduction of the water tension from 72 to 27 mN/m (Bezerra et al. 2019; Rocha e Silva et al. 2019). Surfactin can be produced by the nonpathogenic *Bacillus subtilis*, and similar structures produced are also reported from *Bacillus licheniformis*, with promising stability studies under adverse environmental conditions (Meena and Kanwar 2015).

7.5 Production of Biosurfactants for the Food Industry

The production of biosurfactants by microorganisms and their composition/structure are generally influenced by environmental conditions and the composition of the culture medium, which are decisive for increasing the solubility of insoluble compounds in water and consequent transport of microorganisms to the cell producers. An important factor is the fact that, as the production of biosurfactants is not always conducted through main biosynthetic pathways, it is necessary to use water-insoluble substrates to induce the formation of these secondary metabolites, such as, for example, oils, fats, and/or solid and liquid hydrocarbons (García-Reyes et al. 2018).

Thus, one of the main requirements for the biosurfactant production by microorganisms, such as bacteria and yeasts, is the ability to grow in a medium that contains both a hydrophobic (vegetable oils or hydrocarbons) and hydrophilic (carbohydrates or glycerol) carbon source, to provide the nutritional balance, so that it becomes possible to vary the cultivation conditions to obtain biosurfactants according to the intended application (Clements et al. 2019; Patowary et al. 2019). The synthesis of the polar portion of biosurfactants results from the use of both the hydrophilic substrate via the glycolytic pathway, and via the gluconeogenesis pathway, when a hydrophobic substrate is used. The nonpolar fraction is obtained both from a hydrophilic substrate, through the oxidation of glucose to pyruvate and then conversion to acetyl-CoA, precursor of fatty acid synthesis, and from the lipolytic pathway, when hydrophobic substrates are present in the medium. It is also possible to use, in the same production medium, a water-soluble and an insoluble substrate to produce biosurfactants (Ribeiro et al. 2020a).

In addition to the type of carbon source, each microorganism used also determines the structure of the biosurfactant produced, the majority being generally formed by a mannoprotein or glycolipid structure, or a carbohydrate-lipid-protein complex. Furthermore, depending on the metabolic pathway employed by the microorganism, the substrate conversion rate can be increased, resulting in increased productivity (Elshafie et al. 2015).

7.5.1 Agro-Industrial Wastes as Alternatives for Production

The selection of the substrate depends on the choice of raw material with an adequate balance of nutrients for growth and production, being necessary to contain a high content of carbohydrates and lipids, essential elements to produce biosurfactants. However, the costs of these substrates are high and limit the application of biosurfactants on a commercial scale (Santos et al. 2016; Sindhu et al. 2020). On the other hand, the use of residues as raw material in bioprocesses is a good alternative, if there is a good balance of nutrients to obtain the desired product, so that the production of biosurfactants becomes potentially viable (Moshtagh et al. 2019). Then, new low-cost production strategies have been studied in biotechnological processes aiming at the formulation of alternative media composed of waste residues, since these account for about 50% of the final cost of the product (Lima et al. 2017).

Among the agro-industrial by-products widely produced and used by microorganisms for the production of biosurfactants, apple and clarified cashew juice (Oliveira and Garcia-Cruz 2013), cassava wastewater (Barros et al. 2008), animal fat (Santos et al. 2013), vinasse (Oliveira et al. 2013), vegetable fat (Gusmão et al. 2010), waste frying oil (Campos et al. 2014; Silva et al. 2019), sugarcane molasses (Santos et al. 2017), and corn steep liquor (Soares da Silva et al. 2017) stand out. These residues can be used as carbon and nitrogen sources, replacing raw materials with higher added value. Thus, biosurfactants produced from waste are promising for application not only in the food sector, but also in cosmetics, agrochemicals, and pharmaceuticals (Jiménez-Peñalver et al. 2019).

7.5.2 Biosurfactant-Producing Yeasts for Food Application

Yeasts have shown great potential to produce biosurfactants, as they are industrially widely used in the food sector. This is due to the fact that some are included in GRAS (Generally Regarded as Safe) status, and do not have toxicity and pathogenicity risks (Campos et al. 2015). Some yeasts included in this status are of the *Candida*, *Starmerella*, and *Saccharomyces* genus, being reported by several authors to produce biosurfactants with potential food application. Examples of biosurfactant producers include *Candida lipolytica* (Sarubbo et al. 1999, 2001), *Candida utilis*

(Campos et al. 2019), *Candida bombicola* (Silva et al. 2020), *Starmerella bombicola* (Wang et al. 2019), and *Saccharomyces cerevisiae* (Ribeiro et al. 2020b), known for producing biosurfactants with good results in emulsifying action, antimicrobial, and antioxidant activity.

As for food applications with *C. lipolytica*, Cirigliano and Carman (1984, 1985) were the pioneers in studying the production of a bioemulsifier in a medium composed of several hydrophobic carbon sources. A few years later, it was shown that this yeast can use babassu oil or glucose in a batch and fed batch culture medium (Sarubbo et al. 1999, 2001), being a precursor of studies with canola oil, concluding that the biosurfactant is formed by a complex structure composed of protein, lipid, and polysaccharide, with properties similar to commercial emulsifiers (Sarubbo et al. 2007).

Regarding the biosurfactant production by *C. utilis*, the first results were reported with application in salad creams (Shepherd et al. 1995). Later, the production of a biosurfactant in a medium composed of glucose and canola waste frying oil was described, showing stability in emulsifying action and in surface-active activity after 24 h of observation (Campos et al. 2019).

In relation to *C. bombicola*, several studies have already demonstrated the possibility of biosurfactant production both by glucose and by industrial residues as substrates, such as sugarcane molasses and frying oil, as a carbon source, as well as corn liquor as a nitrogen source, obtaining nontoxic biosurfactants with good emulsifying and antimicrobial activity, with yields around 40 g/L (Minucelli et al. 2017; Resende et al. 2019). As for *S. bombicola*, teleomorph of *C. bombicola* (Rosa and Lachance 1998), studies report the use of agro-industrial residues and high concentration of hydrophobic substrates to produce large amounts of sophorolipids (about 177 g/L) with emulsifying properties (Shah et al. 2017; Lodens et al. 2020).

Regarding biosurfactants produced by *Saccharomyces cerevisiae*, these metabolites are known to have an effective emulsifying property. Since the 1980s, results of emulsifying activity and stability at different temperatures of intracellular mannoproteins extracted from the cell wall of these yeasts have been reported (Cameron et al. 1988). More recent studies also describe the possibility of producing extracellular biosurfactants with better emulsifying activity from culture media supplemented with glucose and oil as carbon source (Hussain Ali and Shawkat Ali 2019). By cultivating this yeast in a medium composed only of soy frying oil and corn steep liquor, it was possible to produce a glycolipid with thermal stability and emulsifying action in cookies, as shown in Fig. 7.3 (Galdino Ribeiro et al. 2020; Ribeiro et al. 2020b).

Fig. 7.3 Cookies added with *S. cerevisiae* biosurfactant at a concentration of 4% in total replacement of the animal fat source (egg yolk)

7.6 Chemical Additives vs. Biosurfactants in the Food Industry

It is true that, despite the limitations regarding the entry of biosurfactants on the market, they advantageously differ from surfactants and chemical additives mainly due to their high biodegradability, an extremely relevant property in all areas of application due to the increase in concern for the environment. In the food industry, this factor, combined with the low toxic potential, gives biosurfactants a benefit due to the consumers' need for foods that cause less or no risk to health and well-being, in contrast to foods that contain substances that may perhaps cause some harm to human health in the long term (Ribeiro et al. 2020a).

Other significant properties of biosurfactants for the food industry include their ability to resist variations in temperature, acidity, and salt concentration, so that they can be applied in formulations that undergo different processes without changing their essential characteristics for the respective applications. Some yeast biosurfactants are examples of molecules that have the property of temperature resistance and can be subjected to conditions above 200 °C without significant loss of mass, according to the results of thermogravimetric analysis (Ribeiro et al. 2020c). In addition, the antioxidant, antiadhesive, and antimicrobial action found for some biosurfactants expands the application possibilities in this promising sector, giving them greater added value compared to chemical additives (Zouari et al. 2016a).

7.7 Application of Biosurfactants in Food

The food industries have always shown concern with the demands and needs of the market and allied to the growing demand for more natural products; over the years, one of the measures was to develop new formulations with the addition of additives from vegetables, which currently have vast application and commercial acceptance, such as soy lecithin and gum arabic, to replace artificial or chemically synthesized compounds (Hasenhuettl and Hartel 2019). However, there are some limitations with the use of these emulsifiers under microwave cooking conditions and irradiation, due to the likelihood of loss of important properties, which increased the interest in producing, isolating, and identifying new compounds more suitable for the intended purposes (Nitschke and Costa 2007; Faustino et al. 2019).

As already mentioned, biosurfactants for food applications have been studied since the 1980s, showing, over the years, that these biomolecules can contribute to several purposes, whether to intensify the aroma through the control of consistency and solubilization of flavoring oils or act as fat stabilizers during the cooking process in meat products (Vijayakumar and Saravanan 2015; Giri et al. 2017). In addition, due to their emulsifying action, they can also control viscosity in foods containing citric and ascorbic acids and improve organoleptic properties in butter creams, breads, and confectionery products. Thus, the food sector has received much attention to expand the application of biosurfactants (Nitschke and Costa 2007; Kieliszek et al. 2017).

7.7.1 Biosurfactants with Antimicrobial and Antiadhesive Action

Some biosurfactants can significantly contribute to the overall quality and safety of foods due to their ability to act as antimicrobial and antiadhesive agents. Thus, they become interesting alternatives to control the proliferation of undesirable microorganisms by preventing their adherence to equipment and utensils used in dairy processing, for example, since the occurrence of microorganism incrustations in heat exchangers is frequent. Thus, by acting as an antiadhesive biological coating, they favor an increase in the shelf life of processed products (Marcelino et al. 2020).

The biosurfactant produced by *Lactobacillus plantarum* demonstrates that it is possible to obtain satisfactory results in the agar well diffusion method at concentrations up to 25 mg/mL, since it has an antiadhesive action greater than 50% against different unwanted microorganisms for foods, i.e., *Escherichia coli* MTCC 108, *E. coli* ATCC 31705, *Yersinia enterocolitica* MTCC 859, *Salmonella typhi* and *Staphylococcus aureus* F 722, in addition to maximum antimicrobial activity of *E. coli* ATCC and *Y. enterocolitica* (Madhu and Prapulla 2014). By the serial microdilution method, 2.6 mg/mL of glycolipid (crude extract) produced by the yeast *Wickerhamomyces anomalus* CCMA 0358 presented an antimicrobial activity of 66% for *Staphylococcus epidermidis*, 78% for *Candida albicans*, and 100% for *Streptococcus agalactiae* (Souza et al. 2017). Marine yeasts have also shown antimicrobial activity, such as *Cyberlindnera saturnus* SBPN-27, which presented an inhibitory action between 69% and 100% for different pathogenic bacteria at a concentration of 200 mg/mL (Balan et al. 2019).

7.7.2 Biosurfactants with Antioxidant Action

Antioxidants are compounds responsible for delaying the appearance of significant changes in foods, preserving them from deterioration, rancidity, and discoloration arising from autoxidation. They can be synthetic, such as butylhydroxyanisole (BHA) and butylhydroxytoluene (BHT), or natural, such as organosulfur, phenolic, and terpenes. Biosurfactants can be included together with natural antioxidants, also contributing to reduce the risk of heart disease and combat degenerative diseases (Hameed et al. 2020; Xu et al. 2021).

Antioxidants can act in different ways, whether in radical scavenging or complex reduction, and biosurfactants can show good results for all or some methods depending on the chemical structure, with the unsaturated fatty acids in the chain being crucial for a high activity. Biosurfactant produced by *Candida utilis*, at a concentration of 5 mg/mL, showed greater capacity to reduce complexes (greater than 70% inhibition) than to scavenge DPPH (2,2-diphenyl-1-picryl-hydrazyl) radicals (Kiran et al. 2017; Ribeiro et al. 2020c). In contrast, at the same concentration, *Lactobacillus* biosurfactants acted better as scavengers of DPPH molecules, with activity above 70% (Merghni et al. 2017).

7.7.3 Biosurfactants with Emulsifying and Stabilizing Action

Emulsifiers, which are one of the most used types of food additives, are compounds used to provide greater creaminess, stability, and yield of food formulations by maintaining the uniform dispersion of one liquid in another, such as oil and water. They play a decisive role in the formation of successful emulsions by facilitating the initial formation of fine lipid droplets during homogenization and increasing their stability (Fig. 7.4) (Mcclements and Gumus 2016). Thus, it is possible to control the

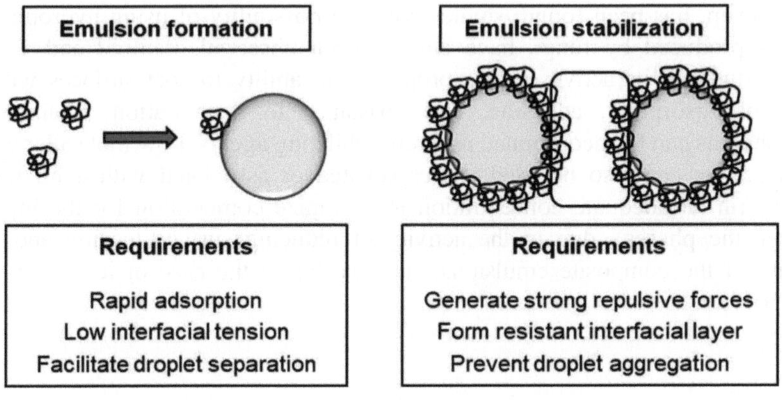

Fig. 7.4 Requirements for the formation and stabilization of emulsions

grouping of globules and aerodynamic systems, promoting the stabilization of phases in heterogeneous systems, very common in the food sector. Furthermore, with its use, the shelf life of the products is sufficiently increased, as well as minimizing phase separation without adding other additives (Pascual-Villalobos et al. 2019).

Therefore, biosurfactants are an encouraging alternative, not only for their greater resistance to adverse conditions but also for allowing the formation of stereostatic and electrostatic barriers, preventing particle coalescence, thus stabilizing emulsions (Mcclements and Gumus 2016). Glycolipid biosurfactants generally act as good emulsifiers in oil-water systems, despite their low molecular weight. Some authors have used microbial surfactants in some foods as effective emulsifying agents, improving consistency and texture (Egolf et al. 2019; Nitschke and Silva 2017). Despite few reports on the use of microbial surfactants as food emulsifiers, the results so far point to significant improvements in foods formulated with these biomolecules, which are interchangeably called bioemulsifiers, whose properties do not necessarily encompass the ability to reduce the medium surface tension (Campos et al. 2019; Gallo et al. 2019).

7.7.3.1 Biosurfactants in Ice Creams

Due to the ability of biosurfactants to control the consistency of products and add flavor due to the ease with which they solubilize aromas, they can be incorporated into ice cream as emulsifying agents, so that physical properties such as creaminess and texture are preserved, contributing to the quality of the final product (Anjum et al. 2016; Kieliszek et al. 2017). In this type of formulation, biosurfactants must only partially prevent the penetration of fat crystals into liquid droplets, through the formation of thin layers, so that partial coalescence occurs, which is desirable during ice cream production (Mcclements and Gumus 2016; Berton-Carabin and Schroën 2019).

Although no concrete evidence of the use of biosurfactants in aerated foods, such as ice cream, has been found, studies with the possibility of using hydrophobins, proteins produced by fungi, have already been observed. Penfold and Thomas (2019), studying its active surface property and ability to coat surfaces with the action of adsorption, adhesion, and resistance to denaturation, realized that hydrophobins can be incorporated as foam stabilizing agents. Low molecular weight biosurfactants can also be used, either isolated or associated with a mixture of proteins (in an adequate concentration to minimize competition for the interface between the phases), due to the activity of reducing surface tension and rapid stability of the composite emulsions, such as this in the case of ice cream (Zhu et al. 2019).

7.7.3.2 Salad Dressings with Added Biosurfactants

The addition of emulsifiers, followed by stirring, in salad dressings interferes with their creamy appearance by leading to the formation of a water-oil type dispersion. Mayonnaise, for example, characterized as an emulsion of vegetable oil in water (vinegar), needs emulsifiers that influence the size of the oil droplets and the proximity with which they are grouped for a more rigid emulsion. For this purpose, the positive contribution of a *Candida utilis* biosurfactant at a concentration of 0.7% (w/v) as an effective emulsifier was reported, improving physical aspects of sauces containing guar gum and carboxymethylcellulose (CMC), and providing better stability after 30 days of storage (Campos et al. 2019). Figure 7.5 shows an example of application of *C. utilis* biosurfactant as emulsifier and stabilizer in mayonnaise type dressing.

Comparing the emulsifying and stabilizing action of biosurfactants produced by yeasts with other commercial emulsifiers, such as gum arabic, guar gum, and carboxymethylcellulose, better results are observed for biosurfactants. This is also due to their greater resistance to heat treatments, so that the granular structure and rheological properties of foods when incorporated with biosurfactants can be better preserved by the stability of their emulsifying action under extreme conditions (Heyman et al. 2014; Ribeiro et al. 2020c).

7.7.3.3 Biosurfactants in Baked Goods and Cookies

Farinaceous foods such as breads, muffuns and cookies are high in demand therefore, these foods serves as great reservoir for biosurfactants as food additives. Biosurfactants can improve the texture of the dough and decrease its energy value (Antoniewska et al. 2018). In these foods, the use of emulsifiers of microbial origin emerged with the aim of reducing the use of emulsifiers currently marketed and improving the rheology of the products, even though there are currently few reports of their use for this purpose (Zouari et al. 2016b).

Fig. 7.5 Example of application of *C. utilis* biosurfactant as emulsifier and stabilizer in mayonnaise-type dressings. (**a**) Carboxymethyl cellulose + biosurfactant. (**b**) Guar gum + biosurfactant. (**c**) Biosurfactant from *C. utilis*

Fig. 7.6 Muffins made from *C. bombicola* biosurfactant replacing 50% of the source of vegetable fat (margarine)

In bakery products, studies with the addition of rhamnolipids and *Bacillus subtilis* biosurfactants showed positive changes in mass, texture profile, and volume in the final formulations. The biosurfactants also reduced the probability of proliferation of unwanted microorganisms (Giri et al. 2017; Mnif et al. 2012). In muffin formulations, there are reports of improved texture profile with reduced hardness and chewability and increased cohesion and elasticity when 0.75% of lipopeptide produced by the halophilic bacterium *Nesterenkonia* sp. is added in the formulation quality (Tess et al. 2015; Kiran et al. 2017). By replacing vegetable fat with *Candida bombicola* URM 3718 biosurfactant, with thermal stability and significant antioxidant activity, the physical and textural properties were preserved, as well as the overall look (Fig. 7.6), in addition to reducing the percentage of trans fats and improving the nutritional value of the final product (Silva et al. 2020).

For biscuit formulations, Zouari et al. (2016b) published a study comparing the addition of *Bacillus subtilis* biosurfactant and a commercial emulsifier GMS (Glycerol Monostearate) in biscuit dough containing wheat flour and sesame, obtaining greater softness and overall quality with the addition of the bioemulsifier. More recently, Ribeiro et al. (2020c), when replacing the percentage of egg yolk by 50% and 100% with glycolipids from *S. cerevisiae* URM 6670 with a health-beneficial fatty acid profile, observed that the physical properties, such as the texture profile of the dough, were maintained when compared to the standard formulation.

7.8 Biosurfactants as Heavy Metal Washing Agents

Heavy metals are chemical substances present throughout the environment, especially in fertilizers and pesticides used in agriculture. When heavy metal are used excessively, they can leave harmful residues in crops and natural environments, which eventually penetrate the human body along food chains, accumulating in the body, causing irreversible damage to organs and their functions (Jimoh and Lin 2019a). Therefore, biosurfactants can contribute to food safety by acting to remove heavy metals present on vegetable surfaces, for example, and can provide healthier and safer food for consumers. Researchers report a significant removal of Cu^{2+}, Fe^{2+},

Ni^{2+}, and Zn^{2+} present in vegetables with the action of a *Paenibacillus* sp. D9 (Jimoh and Lin 2019b). There are also reports of the use of recombinant *Bacillus* sp. MTCC 5877 to optimize the washing action, obtaining cadmium removal percentages of 47–73% from different vegetables (Anjum et al. 2016). By evaluating the removal of this same metal from vegetables such as ginger, carrots, radishes, and potatoes, it was possible to obtain removal of up to 61% with a recombinant *Bacillus licheniformis* VS16 biosurfactant (Giri et al. 2017). Based on these observations, it is possible to expand the use of biosurfactants in various sectors of the food industry.

7.9 Biosurfactants of Application Challenges in the Food Industry

It is true that biosurfactants present themselves as promising molecules for the food industry. However, their industrial use has faced some obstacles such as the economic limitation of its production, since their market cost is still high when compared to the production of synthetic surfactants. To produce a biosurfactant, the cost of the raw material represents about 10–30% of the total cost, and the availability of an adequate method of production and economic recovery associated with a good yield and productivity of the producing microorganism is not yet well established (Radzuan et al. 2017). In addition, considering that the food industry uses processes with changing temperature, pressure, viscosity, pH, and solubility together with microwave, ultrasound, and homogenization technologies for several matrices composed of different proportions of carbohydrates, lipids, proteins, fibers, minerals, and vitamins, it becomes a challenge to reach maximum efficiency of action of biosurfactants in these formulations with the lowest possible application concentration (Augusto 2020). However, consumer preference for a more responsible diet, combined with growing concern about the origin of the ingredients used, may facilitate the use of biosurfactants in the coming years in response to the demand for green food additives (Ribeiro et al. 2020c).

7.10 Conclusion and Future Perspectives

Biosurfactants have become target molecules for applications in several sectors, due to their advantageous characteristics compared to the synthetic surfactants currently used. In the food sector, their potential to act in the improvement of several important parameters was observed, in particular as emulsifiers and stabilizers, in addition to meeting consumer demand for more natural foods, as it is possible to produce them by safe microorganisms and in safe conditions. Thus, they present themselves as promising alternatives to replace commercial emulsifiers with the addition of other unique properties. However, the application of biosurfactants in food is still limited due to some issues such as the availability of economically viable production technologies, with yields not yet sufficient, and the understanding and obtaining of adequate biosurfactants for each purpose. Thus, many studies still need to be carried out to reach ideal conditions for commercialization, despite the advances demonstrated in applying these biomolecules in recent years. Furthermore, how the interactions of biosurfactants with ingredients in food formulations can affect the sensory response of consumers needs to be evaluated so that they can be used successfully. In general, it can be considered that biosurfactants have a recognizable potential for the benefit of human health and the environment and, based on a greater development of research in the food sector, they may be part of the processes in a more expressive and effective way.

References

Anic I, Apolonia I, Franco P et al (2018) Production of rhamnolipids by integrated foam adsorption in a bioreactor system. AMB Exp 8:122. https://doi.org/10.1186/s13568-018-0651-y

Anjum F, Gautam G, Edgard G et al (2016) Biosurfactant production through *Bacillus* sp. MTCC 5877 and its multifarious applications in food industry. Bioresour Technol 213:262–269. https://doi.org/10.1016/j.biortech.2016.02.091

Antoniewska A, Rutkowska J, Pineda MM et al (2018) Antioxidative, nutritional and sensory properties of muffins with buckwheat flakes and amaranth flour blend partially substituting for wheat flour. LWT - Food Sci Technol 89:217–223. https://doi.org/10.1016/j.lwt.2017.10.039

Arab F, Mulligan CN (2018) An eco-friendly method for heavy metal removal from mine tailings. Environ Sci Pollut Res 25(16):16202–16216. https://doi.org/10.1007/s11356-018-1770-3

Araújo SCS, Silva-Portela RCB et al (2020) MBSP1: a biosurfactant protein derived from a metagenomic library with activity in oil degradation. Sci Rep 10:1340. https://doi.org/10.1038/s41598-020-58330-x

Augusto PE (2020) Challenges, trends and opportunities in food processing. Curr Opin Food Sci 35:72–78. https://doi.org/10.1016/j.cofs.2020.03.005

Bages-Estopa S, White DA, Winterburn JB et al (2018) Production and separation of a trehalolipid biosurfactant. Biochem Eng J 139(15):85–94. https://doi.org/10.1016/j.bej.2018.07.006

Bahia FM, de Almeida GC, de Andrade LP et al (2018) Rhamnolipids production from sucrose by engineered *Saccharomyces cerevisiae*. Sci Rep 8:2905. https://doi.org/10.1038/s41598-018-21230-2

Bakhshi N, Soleimanian-Zada S, Sheikh-Zeinoddin M (2017) Dynamic surface tension measurement for the screening of biosurfactants produced by *Lactobacillus plantarum* subsp. *Plantarum* PTCC 1896. Enzyme Microbiol Technol 101:1–8. https://doi.org/10.1016/j.enzmictec.2017.02.010

Balan SS, Kumar CG, Jayalakshmi S (2019) Physicochemical, structural and biological evaluation of Cybersan (trigalactomargarate), a new glycolipid biosurfactant produced by a marine yeast, *Cyberlindnera saturnus* strain SBPN-27. Process Biochem 80:171–180. https://doi.org/10.1016/j.procbio.2019.02.005

Barros FFC, Quadros CP, Pastore G (2008) Propriedades emulsificantes e estabilidade do biossurfactante produzido por *Bacillussubtilis* em manipueira. CiênTecnol Ali 28:979–985. https://doi.org/10.1590/S0101-20612008000400034

Berton-Carabin C, Schroën K (2019) Towards new food emulsions: designing the interface and beyond. Curr Opin Food Sci 27:74–81. https://doi.org/10.1016/j.cofs.2019.06.006

Bezerra KGO, Gomes UVR, Silva RO et al (2019) The potential application of biosurfactant produced by *Pseudomonas aeruginosa* TGC01 using crude glycerol on the enzymatic hydrolysis of lignocellulosic material. Biodegradation 30(4):351–361. https://doi.org/10.1007/s10532-019-09883-w

Cameron DR, Cooper DG, Neufeld RJ (1988) The mannoprotein of *Saccharomyces cerevisiae* is an effective bioemulsifier. Appl Environ Microbiol 54:1420–1425

Campos JM, Stamford TLM, Sarubbo LA (2014) Production of a bioemulsifier with potential application in the food industry. Appl Biochem Biotechnol 172:3234–3252. https://doi.org/10.1007/s12010-014-0761-1

Campos JM, Stamford TLM, Rufino RD et al (2015) Formulation of mayonnaise with the addition of a bioemulsifier isolated from *Candida utilis*. Toxicol Rep 2:1164–1170. https://doi.org/10.1016/j.toxrep.2015.08.009

Campos JM, Stamford TLM, Sarubbo LA (2019) Characterization and application of a biosurfactant isolated from *Candida utilis* in salad dressings. Biodegradation 30:313–324. https://doi.org/10.1007/s10532-019-09877-8

Chen W, Qu Y, Xu Z et al (2017) Heavy metal (Cu, Cd, Pb, Cr) washing from river sediment using biosurfactant rhamnolipid. Environ Sci Pollut Res 24(19):16344–16350. https://doi.org/10.1007/s11356-017-9272-2

Cirigliano MC, Carman GM (1984) Isolation of a bioemulsifier from *Candida lipolytica*. Appl Environ Microbiol 48:747–750

Cirigliano MC, Carman GM (1985) Purification and characterization of liposan, a bioemulsifier from *Candida lipolytica*. Appl Environ Microbiol 50:846–850

Clements T, Ndlovu T, Khan S et al (2019) Biosurfactants produced by *Serratia* species: classification, biosynthesis, production and application. Appl Microbiol Biotechnol 103:589–602. https://doi.org/10.1007/s00253-018-9520-5

Egolf A, Hartmann C, Siegrist M (2019) When evolution works against the future: disgust's contributions to the acceptance of new food technologies. Risk Anal 39:1546–1559. https://doi.org/10.1111/risa.13279

Ekpenyong MG, Antai SP, Asitok AD et al (2017) Plackett-Burman design and response surface optimization of medium trace nutrients for glycolipopeptide biosurfactant production. Iran Biomed J 21(4):249–260. https://doi.org/10.18869/acadpub.ibj.21.4.249

Elshafie AE, Joshi SJ, Al-Wahaibi YM et al (2015) Sophorolipids production by *Candida bombicola* ATCC 22214 and its potential application in microbial enhanced oil recovery. Front Microbiol 6:1324. https://doi.org/10.3389/fmicb.2015.01324

Fari A, Saeed A (2016) Production and biomedical applications of probiotic biosurfactants. Curr Microbiol 72:489–495. https://doi.org/10.1007/s00284-015-0978-4

Farias CBB, Almeida FCG, Silva IA (2021) Production of green surfactants: market prospects. Electr J Biotechnol 51:28–39. https://doi.org/10.1016/j.ejbt.2021.02.002

Faustino M, Veiga M, Sousa P et al (2019) Agro-food byproducts as a new source of natural food additives. Molecules 24(6):1056. https://doi.org/10.3390/molecules24061056

Fukuoka T, Shinozaki Y, Tsuchiya W et al (2016) Control of enzymatic degradation of biodegradable polymers by treatment with biosurfactants, mannosylerythritol lipids, derived from Pseudozyma spp. yeast strains. Appl Microbiol Biotechnol 100:1733–1741. https://doi.org/10.1007/s00253-015-7068-1

Galdino Ribeiro B, Monteiro Dos Santos M, Amaro da Silva I et al (2020) Study of the biosurfactant production by *Saccharomyces cerevisiae* URM 6670 using agroindustrial waste. Chem Eng Trans 79:61–66. https://doi.org/10.3303/CET2079011

Gallo SB, Brochado T, Brochine L et al (2019) Effect of biosurfactant added in two different oil source diets on lamb performance and ruminal and blood parameters. Livestock Sci 226:66–72. https://doi.org/10.1016/j.livsci.2019.06.006

García-Reyes S, Yáñez-Ocampo G, Wong-Villarreal A et al (2018) Partial characterization of a biosurfactant extracted from *Pseudomonas* sp. B0406 that enhances the solubility of pesticides. Environ Technol 39(20):2622–2631. https://doi.org/10.1080/21622515.2017.1363295

Gaur VK, Regar RK, Dhiman N et al (2019) Biosynthesis and characterization of sophorolipid biosurfactant by *Candida* spp.: application as food emulsifier and antibacterial agent. Bioresour Technol 285:121314. https://doi.org/10.1016/j.biortech.2019.121314

Giri SS, Sen SS, Jun JW et al (2017) Role of *Bacillus licheniformis* VS16-derived biosurfactant in mediating immune responses in Carp Rohu and its application to the food industry. Front Microbiol 8:2–13. https://doi.org/10.3389/fmicb.2017.00514

Gouda RK, Pathak M, Khan MK (2020) A biosurfactant as prospective additive for pool boiling heat transfer enhancement. Int J Heat Mass Trans 150:119292. https://doi.org/10.1016/j.ijheatmasstransfer.2019.119292

Gusmão CAB, Rufino RD, Sarubbo LA (2010) Laboratory production and characterization of a new biosurfactant from *Candida glabrata* UCP1002 cultivated in vegetable fat waste applied to the removal of hydrophobic contaminant. World J Microbiol Biotechnol 26:1683–1692. https://doi.org/10.1007/s11274-010-0346-2

Hameed A, Hussain SA, Ijaz MU et al (2020) Antioxidant activity of polyphenolic extracts of filamentous fungus *Mucor circinelloides* (WJ11): extraction, characterization and storage stability of food emulsions. Food Biosci 34:100525. https://doi.org/10.1016/j.fbio.2019.100525

Hasenhuettl GL (2019) Synthesis and commercial preparation of food emulsifiers. In: Hasenhuettl GL, Hartel RW (eds) Food emulsifiers and their applications. Springer, Cham, pp 11–39. https://doi.org/10.1007/978-3-030-29187-7

Heyman BB, De Vos WH, Depypere F et al (2014) Guar and xanthan gum differentially affect shear induced breakdown of native waxy maize starch. Food Hydrocolloids 35:546–556. https://doi.org/10.1016/j.foodhyd.2013.07.011

Hussain Ali L, Shawkat Ali W (2019) Production and antibacterial activity of biosurfactant from *Saccharomyces cerevisiae*. J Phys Confer Ser 1234:012080. https://doi.org/10.1088/1742-6596/1234/1/012080

Irorere VU, Tripathi L, Marchant R et al (2017) Microbial rhamnolipid production: a critical re-evaluation of published data and suggested future publication criteria. Appl Microbiol Biotechnol 101:3941–3951. https://doi.org/10.1007/s00253-017-8262-0

Jiménez-Peñalver P, Koh A, Gross R et al (2019) Biosurfactants from waste: structures and interfacial properties of sophorolipids produced from a residual oil cake. J Surf Deterg 23(2). https://doi.org/10.1002/jsde.12366

Jimoh AA, Lin J (2019a) Biosurfactant: a new frontier for greener technology and environmental sustainability. Ecotoxicol Environ Saf 184:109607. https://doi.org/10.1016/j.ecoenv.2019.109607

Jimoh AA, Lin J (2019b) Enhancement of *Paenibacillus* sp. D9 lipopeptide biosurfactant production through the optimization of medium composition and its application for biodegradation of hydrophobic pollutants. Appl Biochem Biotechnol 187:724–743. https://doi.org/10.1007/s12010-018-2847-7

Kaczorek E, Pacholak A, Zdarta A et al (2018) The impact of biosurfactants on microbial cell properties leading to hydrocarbon bioavailability increase. Colloids Interfaces 2(3):35. https://doi.org/10.3390/colloids2030035

Karlapudi AP, Venkateswarulu TC, Tammineedi J et al (2018) Role of biosurfactants in bioremediation of oil pollution—a review. Petroleum 4:241–249. https://doi.org/10.1016/j.petlm.2018.03.007

Kieliszek M, Kot AM, Bzducha-Wróbel A et al (2017) Biotechnological use of *Candida* yeasts in the food industry: a review. Fungal Biol Rev 31:185–198. https://doi.org/10.1016/j.fbr.2017.06.001

Kiran GS, Priyadharsini S, Sajayan A et al (2017) Production of lipopeptide biosurfactant by a marine *Nesterenkonia* sp. and its application in food industry. Front Microbiol 8:1138. https://doi.org/10.3389/fmicb.2017.01138

Kuyukina MS, Ivshina IB (2019) Production of trehalolipid biosurfactants by *Rhodococcus*. In: Alvarez HM (ed) Biology of Rhodococcus, Microbiology monographs, vol 16. Springer, Berlin, pp 271–298. https://doi.org/10.1007/978-3-030-11461-9_10

Lima RA, Andrade RFS, Rodríguez DM et al (2017) Production and characterization of biosurfactant isolated from *Candida glabrata* using renewable substrates. Afr J Microbiol Res 11(6):237–244. https://doi.org/10.5897/ajmr2016.8341

Lima MT, Groß M, Kurt-Zerdeli SN et al (2019) Surface and aggregation properties of a plant-oil derived biosurfactant. Colloids Surf B Biointerfaces 174:521–527. https://doi.org/10.1016/j.colsurfb.2018.11.055

Liu Y, Zeng G, Zhong H et al (2017) Effect of rhamnolipid solubilization on hexadecane bioavailability: enhancement or reduction? J Hazard Mater 322:394–401. https://doi.org/10.1016/j.jhazmat.2016.10.025

Liu K, Sun Y, Cao M et al (2019) Rational design, properties and applications of biosurfactants: a short review of recent advances. Curr Opin Colloids Interface Sci 45:57–67. https://doi.org/10.1016/j.cocis.2019.12.005

Lodens S, Roelants SLKW, Ciesielska K et al (2020) Unravelling and resolving inefficient glucolipid biosurfactants production through quantitative multiomics analyses of *Starmerella bombicola* strains. Biotechnol Bioeng 117:453–465. https://doi.org/10.1002/bit.27191

Maddikeri GL, Gogate PR, Pandit AB (2015) Improved synthesis of sophorolipids from waste cooking oil using fed batch approach in the presence of ultrasound. Chem Eng J 263:479–487. https://doi.org/10.1016/j.cej.2014.11.010

Madhu AN, Prapulla SG (2014) Evaluation and functional characterization of a biosurfactant produced by *Lactobacillus plantarum* CFR 2194. Appl Biochem Biotechnol 172:1777–1789. https://doi.org/10.1007/s12010-013-0649-5

Marangon CA, Martins VCA, Ling MH (2020) Combination of rhamnolipid and chitosan in nanoparticles boosts their antimicrobial efficacy. ACS Appl Mater Interfaces 12(5): 5488–5499. https://doi.org/10.1021/acsami.9b19253

Marcelino PRF, Gonçalves F, Jimenez IM et al (2020) Sustainable production of biosurfactants and their applications. In: Chandel AP, Silva SS (eds) Lignocellulosic biorefining technologies. John Wiley & Sons Ltd, New York. https://doi.org/10.1002/9781119568858.ch8

Mcclements DJ, Gumus CE (2016) Natural emulsifiers—biosurfactants, phospholipids, biopolymers, and colloidal particles: molecular and physicochemical basis of functional performance. Adv Colloids Interface Sci 234:3–26. https://doi.org/10.1016/j.cis.2016.03.002

Meena KR, Kanwar SS (2015) Lipopeptides as the antifungal and antibacterial agents: applications in food safety and therapeutics. Biomed Res Int 2015:473050. https://doi.org/10.1155/2015/473050

Merghni A, Dallel I, Noumi E et al (2017) Antioxidant and antiproliferative potential of biosurfactants isolated from *Lactobacillus casei* and their anti-biofilm effect in oral *Staphylococcus aureus* strains. Microb Pathog 104:84–89. https://doi.org/10.1016/j.micpath.2017.01.017

Minucelli T, Ribeiro-Viana RM, Borsato D et al (2017) Sophorolipids production by *Candida bombicola* ATCC 22214 and its potential application in soil bioremediation. Waste Biomass Valor 8(3):743–753. https://doi.org/10.1007/s12649-016-9592-3

Mnif I, Ghribi D (2016) Glycolipid biosurfactants: main properties and potential applications in agriculture and food industry. J Sci Food Agric 96(13):4310–4320. https://doi.org/10.1002/jsfa.7759

Mnif I, Besbes S, Ellouze R et al (2012) Improvement of bread quality and bread shelf-life by *Bacillus subtilis* biosurfactant addition. Food Sci Biotechnol 21:1105–1112. https://doi.org/10.1007/s10068-012-0144-8

Mnif I, EllouzChaabouni S, Ghribi D (2017) Glycolipid biosurfactants, main classes, functional properties and related potential applications in environmental biotechnology. J Polym Environ 4:1–15. https://doi.org/10.1007/s10924-017-1076-4

Moshtagh B, Hawboldt K, Zhang B (2019) Optimization of biosurfactant production by *Bacillus subtilis* N3-1P using the brewery waste as the carbon source. Environ Technol 40(25):1–10. https://doi.org/10.1080/09593330.2018.1473502

Najmi Z, Ebrahimipour G, Franzetti A et al (2018) In situ downstream strategies for cost-effective bio/surfactant recovery. Biotechnol Appl Biochem 65(4):523–532. https://doi.org/10.1002/bab.1641

Nitschke M, Costa SGVAO (2007) Biosurfactants in food industry. Trends Food Sci Technol 18: 252–259. https://doi.org/10.1016/j.tifs.2007.01.002

Nitschke M, Silva SSE (2017) Recent food applications of microbial surfactants. Crit Rev Food Sci Nutr 58(4):631–638. https://doi.org/10.1080/10408398.2016.1208635

Ohadi M, Forootanfar H, Dehghan-Noudeh G et al (2019) Antimicrobial, anti-biofilm, and anti-proliferative activities of lipopeptide biosurfactant produced by *Acinetobacter junii* B6. Microb Pathog 138:103806. https://doi.org/10.1016/j.micpath.2019.103806

Oliveira JG, Garcia-Cruz CH (2013) Properties of a biosurfactant produced by *Bacillus* pumilus using vinasse and waste frying oil as alternative carbon sources. Braz Arch Biol Technol 56: 155–160. https://doi.org/10.1590/S1516-89132013000100020

Oliveira DWF, Franc IWL, Félix AKN et al (2013) Kinetic study of biosurfactant production by *Bacillus subtilis* LAMI005 grown in clarified cashew apple juice. Colloids Surf B Biointerfaces 101:34–43. https://doi.org/10.1016/j.colsurfb.2012.06.011

Ozdener MH, Ashby RD, Jyotaki M et al (2019) Sophorolipid biosurfactants activate taste receptor type 1 member 3-mediated taste responses and block responses to bitter taste in vitro and in vivo. J Phys Chem B 123(17):3841. https://doi.org/10.1021/acs.jpcb.9b01013

Pascual-Villalobos MJ, Guirao P, Díaz-Baños FG et al (2019) Oil in water nanoemulsion formulations of botanical active substances. In: Nano-biopesticides today and future perspectives. Academic, New York, pp 223–247. https://doi.org/10.1016/b978-0-12-815829-6.00009-7

Patowary K, Das M, Patowary R et al (2019) Recycling of bakery waste as an alternative carbon source for rhamnolipid biosurfactant production. Surf Deterg 22(2):373–384. https://doi.org/10.1002/jsde.12242

Penfold J, Thomas RK (2019) Adsorption properties of plant based bio-surfactants: insights from neutron scattering techniques. Adv Colloids Interface Sci 274:102041. https://doi.org/10.1016/j.cis.2019.102041

Perfumo A, Rudden M, Marchant R et al (2017) Biodiversity of biosurfactants and roles in enhancing the (bio)availability of hydrophobic substrates. In: Krell T (ed) Cellular ecophysiology of microbe, handbook of hydrocarbon and lipid microbiology. Springer, Berlin. https://doi.org/10.1007/978-3-319-20796-4_35-1

Radzuan MN, Banat IM, Winterburn J (2017) Production and characterization of rhamnolipid using palm oil agricultural refinery waste. Bioresour Technol 225:99–105. https://doi.org/10.1016/j.biortech.2016.11.052

Resende AHM, Farias JM, Silva DDB et al (2019) Application of biosurfactants and chitosan in toothpaste formulation. Colloids Surf B Biointerfaces 181:77–84. https://doi.org/10.1016/j.colsurfb.2019.05.032

Ribeiro BG, Guerra JMC, Sarubbo LA (2020a) Biosurfactants: production and application prospects in the food industry. Biotechnol Prog 36:e3030. https://doi.org/10.1002/btpr.3030

Ribeiro BG, Guerra JMC, Sarubbo LA (2020b) Potential food application of a biosurfactant produced by *Saccharomyces cerevisiae* URM 6670. Front Bioeng Biotechnol 8:434. https://doi.org/10.3389/fbioe.2020.00434

Ribeiro BG, Veras BO, Aguiar JS et al (2020c) Biosurfactant produced by *Candida utilis* UFPEDA1009 with potential application in cookie formulation. Electr J Biotechnol 46:14–21. https://doi.org/10.1016/j.ejbt.2020.05.001

Rocha e Silva FCP, Rocha e Silva NMP, Luna JM et al (2018) Dissolved air flotation combined to biosurfactants: a clean and efficient alternative to treat industrial oily water. Rev Environ Sci Bio/Technol 17(4):591–602. https://doi.org/10.1007/s11157-018-9477-y

Rocha e Silva NMP, Meira HM, Almeida FCG et al (2019) Natural surfactants and their applications for heavy oil removal in industry. Sep Purif Rev 48:267–281. https://doi.org/10.1080/15422119.2018.1474477

Rosa CA, Lachance MA (1998) The yeast genus Starmerella gen. nov. and Starmerella bombicola sp. nov., the teleomorph of *Candida bombicola* (Spencer, Gorin & Tullock) Meyer & Yarrow. Int J Syst Bacteriol 48(4):1413–1417. https://doi.org/10.1099/00207713-48-4-1413

Sales da Silva IG, Almeida FCG, Rocha e Silva NMP et al (2020) Soil bioremediation: overview of technologies and trends. Energies 13:4664. https://doi.org/10.3390/en13184664

Santos DKF, Rufino RD, Luna JM et al (2013) Synthesis and evaluation of biosurfactant produced by *Candida lipolytica* using animal fat and corn steep liquor. J Petrol Sci Eng 105:4–50. https://doi.org/10.1016/j.petrol.2013.03.028

Santos DKF, Luna JM, Rufino RD et al (2016) Biosurfactants: multifunctional biomolecules of the 21st century. Int J Mol Sci 17:401–430. https://doi.org/10.3390/ijms17030401

Santos DKF, Meira HM, Rufino RD et al (2017) Biosurfactant production from *Candida lipolytica* in bioreactor and evaluation of its toxicity for application as a bioremediation agent. Process Biochem 8:1–11. https://doi.org/10.1016/j.procbio.2016.12.020

Santos APP, Silva MDS, Costa EVL et al (2018) Production and characterization of a biosurfactant produced by *Streptomyces* sp. DPUA 1559 isolated from lichens of the Amazon region. Braz J Med Biol Res 51(2):e6657. https://doi.org/10.1590/1414-431x20176657

Sarubbo LA, Marçal MCR, Porto ALF et al (1999) The use of babassu oil as substrate to produce bioemulsifiers by *Candida lipolytica*. Can J Microbiol 45(5):423–426. https://doi.org/10.1139/w99-025

Sarubbo LA, Marçal MC, Neves MLC (2001) Bioemulsifier production in batch culture using glucose as carbon source by *Candida lipolytica*. Appl Biochem Biotechnol 95:59–67. https://doi.org/10.1385/abab:95:1:59

Sarubbo LA, Farias CBB, Campos-Takaki GM (2007) Co-utilization of canola oil and glucose on the production of a surfactant by *Candida lipolytica*. Curr Microbiol 54(1):68–73. https://doi.org/10.1007/s00284-006-0412-z

Shah MUH, Sivapragasam M, Moniruzzaman M et al (2017) Production of sophorolipids by *Starmerella bombicola* yeast using new hydrophobic substrates. Biochem Eng J 127:60–67. https://doi.org/10.1016/j.bej.2017.08.005

Shakeri F, Babavalian H, Amoozegar MA et al (2021) Production and application of biosurfactants in biotechnology. Biointerface Res Appl Chem 11(3):10446–10460. https://doi.org/10.33263/BRIAC113.1044610460

Sharma D, Saharan BS (2016) Functional characterization of biomedical potential of biosurfactant produced by *Lactobacillus helveticus*. Biotechnol Rep 11:27–35. https://doi.org/10.1016/j.btre.2016.05.001

Shepherd R, Rockey J, Sutherland IW (1995) Novel bioemulsifiers from microorganisms for use in foods. J Biotechnol 40(3):207–217. https://doi.org/10.1016/0168-1656(95)00053-S

Shreve GS, Makula R (2019) Characterization of a new rhamnolipid biosurfactant complex from *Pseudomonas* isolate DYNA270. Biomol Ther 9(12):885. https://doi.org/10.3390/biom9120885

Silva MG, Sarubbo LA (2021) Synthetic and biological surfactants used to mitigate biofouling on industrial facilities surfaces. Biointerface Res Appl Chem 12:2560–2585. https://doi.org/10.33263/BRIAC122.25602585

Silva EJ, Silva IA, Brasileiro PPF et al (2019) Treatment of oily effluent using a low-cost biosurfactant in a flotation system. Biodegradation 30:335–350. https://doi.org/10.1007/s10532-019-09881-y

Silva IA, Veras BO, Ribeiro BG et al (2020) Production of cupcake-like dessert containing microbial biosurfactant as an emulsifier. PeerJ 8:e9064. https://doi.org/10.7717/peerj.9064

Sindhu R, Binod P, Nair RB et al (2020) Waste to wealth: valorization of food waste for the production of fuels and chemicals. In: Current developments in biotechnology and bioengineering. Elsevier, Amsterdam, pp 181–197. https://doi.org/10.1016/B978-0-444-64321-6.00009-4

Soares da Silva RCF, Almeida DG, Meira HM et al (2017) Production and characterization of a new biosurfactant from *Pseudomonas cepacia* grown in low-cost fermentative medium and its application in the oil industry. Biocatal Agric Biotechnol 12:206–215. https://doi.org/10.1016/j.bcab.2017.09.004

Solomon OB, Oluwafemi OA, Muktar AG et al (2017) Production and characterization of biosurfactants by *Candida boleticola* H09 and *Rhodotorula bogoriensis* H15 for crude oil recovery and cleaning of oil contaminated fabrics. Int J Life Sci Technol 10(15):109–123

Souza KST, Gudina EJ, Azevedo Z et al (2017) New glycolipid biosurfactants produced by the yeast strain *Wickerhamomyces anomalus* CCMA 0358. Colloids Surf B Biointerfaces 154:373–382. https://doi.org/10.1016/j.colsurfb.2017.03.041

Tess M, Bhaduri S, Ghatak R et al (2015) Physical, textural and sensory characteristics of gluten-free muffins prepared with teff flour (eragrostis tef (zucc) trotter). J Food Process Technol 6(9): 1000490. https://doi.org/10.4172/2157-7110.1000490

Varjani S, Upasani VN (2019) Evaluation of rhamnolipid production by a halotolerant novel strain of *Pseudomonas aeruginosa*. Bioresour Technol 288:121577. https://doi.org/10.1016/j.biortech.2019.121577

Vijayakumar S, Saravanan V (2015) Biosurfactants-types, sources and applications. Res J Microbiol 10(5):181–192. https://doi.org/10.3923/jm.2015.181.192

Wang H, Roelants SLKW, To MH et al (2019) *Starmerella bombicola*: recent advances on sophorolipids production and prospects of waste stream utilization. J Chem Technol Biotechnol 94:999–1007. https://doi.org/10.1002/jctb.5847

Williams W, Trindade M (2017) Metagenomics for the discovery of novel biosurfactants. In: Charles TC et al (eds) Functional metagenomics: tools and applications. Springer, Cham. https://doi.org/10.1007/978-3-319-61510-3_6

Xu X, Liu A, Hu S et al (2021) Synthetic phenolic antioxidants: metabolism, hazards and mechanism of action. Food Chem 353:129488. https://doi.org/10.1016/j.foodchem.2021.129488

Yalaoui-Guellal D, Fella-Temzi S, Djafri-Dib S et al (2020) Biodegradation potential of crude petroleum by hydrocarbonoclastic bacteria isolated from Soummam wadi sediment and chemical-biological proprieties of their biosurfactants. J Petrol Sci Eng 184:106554. https://doi.org/10.1016/j.petrol.2019.106554

Zhong H, Yang X, Tan F et al (2016) Aggregate-based sub-CMC solubilization of n-alkanes by monorhamnolipid biosurfactant. New J Chem 40(3):2028–2035. https://doi.org/10.1039/c5nj02108a

Zhu L, Xu Q, Liu X et al (2019) Oil-water interfacial behavior of soy β-conglycinin–soyasaponin mixtures and their effect on emulsion stability. Food Hydrocolloids 101:105531. https://doi.org/10.1016/j.foodhyd.2019.105531

Zouari R, Moalla-Rekik D, Sahnoun Z et al (2016a) Evaluation of dermal wound healing and in vitro antioxidant efficiency of Bacillus *subtilis* SPB1 biosurfactant. Biomed Pharmacother 84: 878–891. https://doi.org/10.1016/j.biopha.2016.09.08484

Zouari R, Besbes S, Ellouze-Chaabouni S et al (2016b) Cookies from composite wheat–sesame peels flours: dough quality and effect of *Bacillus subtilis* SPB1 biosurfactant addition. Food Chem 194:758–769. https://doi.org/10.1016/j.foodchem.2015.08.064

Chapter 8
Additives in Dairy-Based Food

Harshika Sharma, Gaytri Mahajan, Manpreet Kaur, and Reena Gupta

Abstract Additives have been widely used in food industries since the twentieth century to maintain physical properties like smell, color, texture, etc. and to prevent the food product from microbial attacks. Nowadays, food additives are being used to increase the flavor, texture, emulsification, and stabilization of the food at the time of manufacturing and processing of the product at the industrial level. An additive can be natural or synthetic, depending upon the chemical formula and complexity of the compound. Additives can be categorized as preservatives, colorants, antioxidants, flavoring agents, thickeners, humectants, emulsifying agents, antifoaming agents, stabilizers, etc., which are generally used in dairy-based products. Natural additives are in high demand because some artificial additives have toxic effects on health like cancer, attention deficit disorder, disorders in the immune system, digestive disorders like diarrhea and colicky pains, etc. Because of this reason there is high pressure in increasing the yield of natural additives. In the USA, the guidance about the concentration and toxicity of additive use is regulated by Acceptable Daily Intake (ADI) and Generally Regarded as Safe (GRAS) status while, in India, it is generally guided by the Food Safety and Standards Authority of India (FSSAI). In order to maintain the safety of consumers, the use of additives must be restricted and verified in accordance with national and international laws.

Keywords Additive · Preservative · Antimicrobial · Antioxidant · Thickener · Emulsifying agent · Stabilizer

8.1 Introduction

In the modern food industry, food additives play an important role in maintaining the physical and chemical properties of food like color, smell, taste, and nutrition. Additives are also used to enhance the shelf life of various food products (Wu et al. 2013a, b; Wang et al. 2015). The use of additives is common practice

H. Sharma · G. Mahajan · M. Kaur · R. Gupta (✉)
Department of Biotechnology, Himachal Pradesh University, Shimla, India

A. K. Nadda, G. Goel (eds.), *Microbes for Natural Food Additives*, Microorganisms for Sustainability 38, https://doi.org/10.1007/978-981-19-5711-6_8

on a global scale (Amin et al. 2013; Cai and Liu 2014). In recent years, improved life style of consumers have led to an increase in the demand of natural and nonadditive food products. To overcome the increased demand, synthetic food additives with potential food safety hazards are being introduced to the market. Despite the widespread use of additives, many consumers still believe that additives are just unnecessary to use (Christensen et al. 2011; Chen et al. 2015). Food additives are difficult to understand about because of their complex and diverse functions, which in turn leads to deceptive and misleading information about the additives. Along with this, a number of banned additives or nonedible chemical additives have also been characterized as food additives which are capable of causing health as well as environmental hazards. Technical and accurate information about food additives remains unexplained because the information about the hazards of food additives is dominated, leading consumers to worry about the potential risks caused by food additives (Chen and Han 2009).

8.2 Additives

Additives are those substances which are added in small quantities in food to preserve its flavor or to improve taste, appearance, or other physical and chemical properties. These have been used since ancient times as part of an effort to preserve food. Salt, vinegar, pickling agent, smoke, sugar, etc. are the ancient sources of additives which allow foods like sweets and wines to last longer. In the twentieth century, natural and artificial additives were introduced. The food additives used indirectly to preserve the food at the time of manufacturing, packaging, or storage are known as indirect additives.

A unique number, which is known as E-number, is assigned to every additive for its regulation and ease of use. Codex Alimentarius Commission has extended and adopted the E-numbering design to identify and classify a number of food additives, regardless of their use. For example, products which contain acetic acid are mentioned with a number E260 while in other countries it is simply written as additive 260. Another additive alkannin (additive 103) does not have any E-number, since it is not approved in Europe, although in Australia and New Zealand its use is acceptable.

8.3 Types of Additives

8.3.1 Natural Additives

Natural additives are made up of chemicals found in the environment. These are generally considered as harmless, originate from salt, sugar, vinegar, alcohol, etc. and have high demand in the market. These natural additives still contain some types of impurities. These compounds are also purified in order to remove accompanying substances that do not play a significant role in the final end product. This step is difficult to perform since raw materials cannot be easily differentiated from the natural additives.

8.3.2 Synthetic Additives

Synthetic food additives are synthesized compounds added to the food products to enhance the flavor, texture, appearance, and shelf life of the final product. These are widely produced and used in industrialized countries by chemical and enzymatic methods in order to preserve the food and enhance their properties. A number of synthetic additives are known to be associated with cancer, digestive disorders, and behavioral changes in humans and other animal models. Saccharin is a well-known example of a synthetic additive accepted for edible purposes by the FDA in the United States in early 1907 when additives were considered unsafe for human health.

8.4 Uses of Additives

Additives are added to food for one or many reasons including enhancement of shelf life and flavor enhancement. Food additives are also known to increase the nutritional value of food items. Different uses of additives are depicted in Fig. 8.1.

8.5 Classification of Additives

8.5.1 Preservatives

From ancient time, chemicals have been used to preserve fresh food for later use. This mainly includes drying, cooling, fermenting, and heating to increase the shelf life of edible food items. Several chemical preservatives, namely salt, nitrite, and sulfite, have been used for a number of years, whereas others have been used tremendously. Chemical preservatives have been extensively used to manufacture and market food items in modern times (Davidson and Branen 2005). Today, customers assume food to be easily available, free from foodborne microbes, and to have a fairly longer shelf life (Bacak 2017). Due to the overuse of chemical preservatives, microbes become resistant and there is more demand for natural or nonchemical preservatives in the market (Mahmud and Khan 2018). In general, food

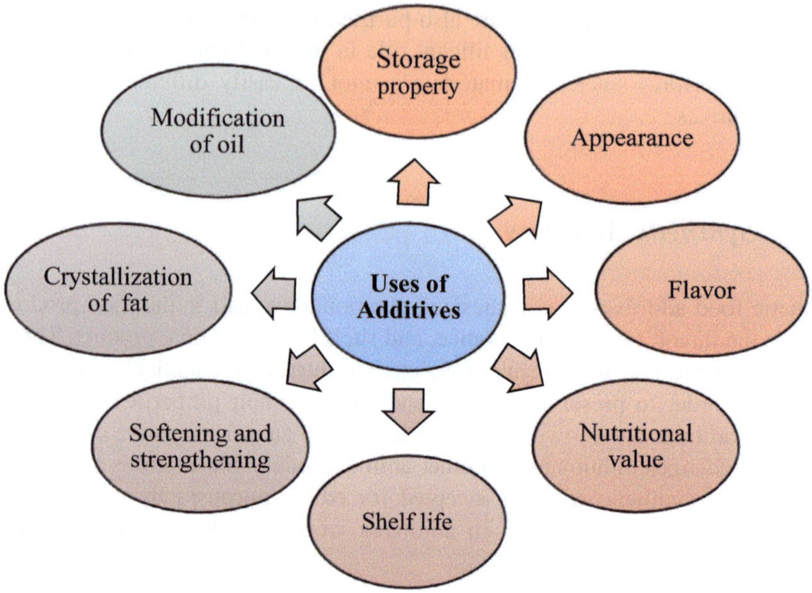

Fig. 8.1 Uses of additives

preservatives are chemical compounds that stimulate food, slow down changes brought about by microbial growth, or prevent the growth of microorganisms from changing the food's physical properties, chemical composition, or nutritional value. The extracts of different compounds and herbal essential oils have inhibitory effects against a number of organisms including insects, bacteria, viruses, and fungi (Xing et al. 2012). Essential oils are derived from clove, oregano, citrus plants and fruits, coriander, garlic, cinnamon, lemongrass, parsley, sage, and rosemary. They act as antioxidant and antimicrobial agents which find applications in bioreactive packaging of food products (Gonelimali et al. 2018). In ancient times, some chemicals have been used as oblique or direct inhibitors to inhibit the growth of microbes and are nevertheless extensively used without considering any restrictions. Preservatives play a key part in food industries, starting from production and dispensation to final customer. The preference of a preservative takes into account the product to be preserved, the kind of microbial spoilage indigenous to it, the product pH, term of shelf life, and simplicity of administration. Using bacteria or their bioreactive metabolites to inhibit foodborne pathogens is bio-preservation (Tumbarski et al. 2020). The microbial peptides and proteins have an advantage that these do not harm or change the quality of food (Rai et al. 2016). As a result, blends or mixtures are frequently used. In a number of food items, particular preservatives do have little or no competition. In general, the concentration used to preserve the food represses the activity of microbes.

8.5.1.1 Characteristics of Food Preservatives

Food preservatives should be simple, economical, and easily available. They should possess acceptable inhibition activity against a variety of microbial pathogens and high stability in food products in which they are added. These preservatives should not change the identity and characteristic attributes of the product and should be practicable and suitable with the processing of final product.

8.5.1.2 Selection of Food Preservatives

Preservatives need to be introduced into the batch quickly after the fruits or vegetables are mashed or the batch is mixed. Any sort of delay results in the growth of microorganisms. Even delay of some hours may allow fermentation to start and give rise to enzymes. These enzymes continue to degenerate the product even after the addition of preservatives. If the additives are water-soluble, it is better to add them to the solution to eliminate the possibility of irregular scattering in food. If the additives are added in powder form, they should be first stable at room temperature. The fermentation might start earlier in these types of solutions. In case of a viscous or thick additive, it is necessary to make a uniform suspension of the additive before its addition into the food product. All the receptacles like bottles, carboys, and barrels should be thoroughly cleaned with steam, hot water, or a 0.02% solution of sodium hydrochloride. Good sanitary practices must be used to get the maximum outcome of preservatives and their conjugation. Food preservatives to be used in a particular food must be decided by some of the major factors as shown in Fig. 8.2.

Majorly the preservatives are divided into two categories, i.e., class I and class II as shown in Table 8.1.

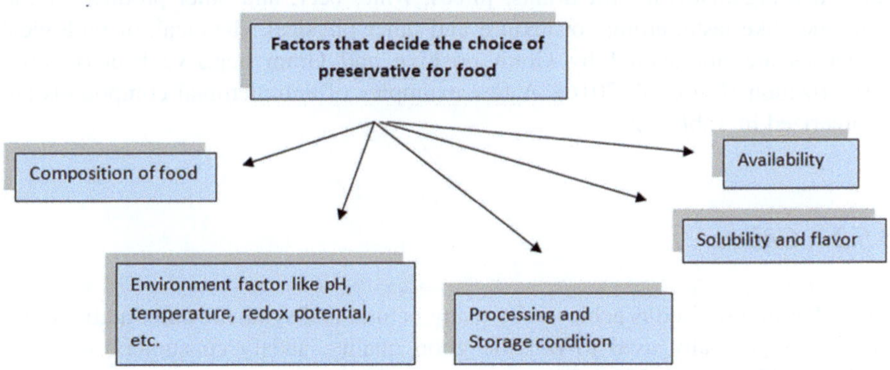

Fig. 8.2 Factors that decide the choice of preservative for food

Table 8.1 Classification of preservatives

Classification of preservatives	
Class I	Class II
Table salt, sugar, dextrose, glucose, spices, vinegar, acetic acid, honey, edible vegetable oils, etc.	Benzoic acid, sulfhurous acid, sorbic acid, propionic acid, lactic acid, sodium diacetate, etc.
There are no restrictions for the addition of class I preservatives to the food items	FSSAI guidelines must be followed for adding class II preservatives to the food items

8.5.1.3 Antimicrobial Agents

Due to the shortage of food, consumers expect food to be available the whole year and that food should be free from foodborne pathogens so that it can have a comparatively longer shelf life (Davidson et al. 2001). Packaging and processing have improved the preservation of food without the use of chemicals.

Antimicrobial products help to reduce and prevent microbial growth. The reduction in water availability and acidic nature is the basic principle/mechanism of antimicrobial agents. Other characteristics of food like flavor, color, texture, and nutritional value are also preserved by these agents. Sorbic acid, potassium and sodium sorbate, calcium and sodium propionates, benzoic acid, etc. are the primary food additives. Growth of bacteria like *B. cereus*, *L. monocytogenes, S. aureus*, *E. coli*, and *Salmonella anatum* is inhibited by the use of cinnamon (Shan et al. 2007). In addition to these compounds, several organic compounds like citric acid, malic acid, lactic acid, and ascorbic acid also act as antimicrobial agents. Lactic acid bacteria (LAB) lead to the production of lactic acid, hydrogen peroxide, fatty acids, acetaldehyde, and benzalkonium chlorides (BACs) that resist microbial growth. Therefore, their cultures are mainly used in dairy food products (Amiri et al. 2020). Sulfur dioxide and sulfites are also extensively used for controlling undesirable microorganisms in soft drinks, juices, wine, beer, and other products. Food properties like taste, aroma, or texture and other physical, chemical, or biological properties are not altered by Gram positive and Gram negative bacteria after pasteurization (Rai et al. 2016). A few examples of antimicrobial compounds are summarized in Table 8.2.

8.5.2 Colors

The coloration of food is achieved by using natural and synthetic color additives on food. It is generally used to provide good quality, satisfy consumers, or assist marketing (Peira et al. 2018; Teigiserova et al. 2020).

"Coloration" does not include any contamination or substance added to food for the purpose of maintaining and improving its nutritional qualities (Codex Alimentarius Commission 2017). Natural food colors are lost by exposure to air,

Table 8.2 Various antimicrobial compounds present in food products

S. no.	Antimicrobial compounds	Target microbes	Food products
1	Acetic acid, acetates, diacetates, dehydroacetic acid	Yeast, bacteria	Dairy products, meats, sauces
2	Lactic acid	Bacteria	Meat and fermented food
3	Lysozyme	Bacteria	Cheese, cooked meat, poultry product
4	Natamycin	Molds	Cheese
5	Nisin	Bacteria	Cheese, cooked meat, poultry product
6	Nitrite, nitrate	Bacteria	Cured meats
7	Propionic acid, propionates	Molds	Bakery products, dairy products

light, temperature, moisture, or improper storage conditions, which can be restored by colorants. These are added to food matrices to obtain the desired food products. It also adds vitamin to the food due to its antioxidant potential. Other than this, it is also used to provide color to colorless food, to protect the flavor, and to enhance the general appearance and nutritional value of food (Benucci et al. 2022). Color additives are of two types, i.e., primary colors and secondary colors (Rodriguez-Amaya 2016). Primary colors are used without dilution whereas secondary colors are blends of primary colors with solvents and other additives (Carmen 2007). Dyes are water-soluble pigments, and they are generally used in beverages, dry mixes, baked goods, dairy products, etc. It is an important indicator of food quality. Good processing and safety of food color are important for marketing. Natural pigments or synthetic organic dyes result in the color of food (de Souza et al. 2021). The basic categories of colorants are depicted in Fig. 8.3.

Colorants or dyes are used as food additives to improve the presentation of the food or to provide color to the food (Sigurdson et al. 2017). Natural pigments are in the form of dye like hydrophilic powders or lipophilic oleoresin. Anthocyanins, carotenoids, betalains, and chlorophyll are some of the most commonly used natural pigments (Miranda et al. 2021). The consumption of these natural colored compounds is used in dietary application because it reduces the risk of diseases like cancer, diabetes (Brennan 2005), and obesity (Cortez et al. 2017). A few examples of natural colorants are mentioned in Table 8.3.

8.5.2.1 Caramels

Caramels have been used in home cooking since ancient times and are responsible for giving dark brown or black products. The basic function of caramels is to provide a specific aroma or color to foods. These are made up of carbohydrates or their combinations like sucrose, glucose, fructose, invert sugar, lactose, maltose, molasses etc. in the presence of acids, alkalis, or salts. According to the Food and Drug Administration (FDA), caramel is the dark brown liquid or solid material that is the result of controlled heat treatment of carbohydrate material. There are four categories

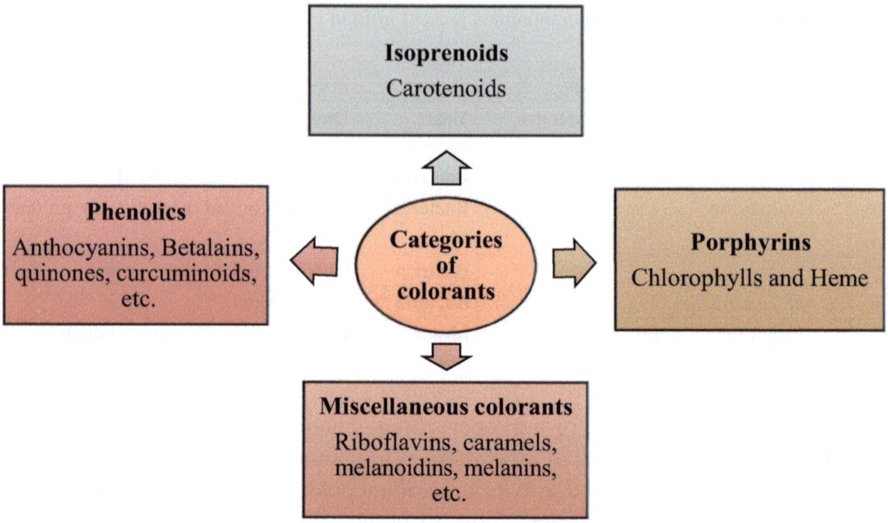

Fig. 8.3 Categories of colorants

Table 8.3 Some common examples of natural colorants

S. no.	Natural colorants	Color produced
1	Caramel	Brown or black
2	Carotenoids	Yellow color
3	Annatto	Orange-red
4	Annatto cheese color	Yellow or orange
5	Annatto butter color	Bright yellow
6	Paprika oleoresin	Orange
7	Saffron	Yellow or orange
8	Carrot extract	Red or purple
9	Chlorophyll	Green
10	Anthocyanin	Blue, red, or purple
11	Beetroot	Red
12	Riboflavin	Yellow or orange

of caramel according to FSSAI (Table 8.4). Caramels can be used as both flavoring and coloring compounds. Food and beverages mostly use caramel color as a food coloring agent.

8.5.2.2 Carotenoids

The source of carotenoids is alfalfa, carrot, tomatoes, citrus peel, palm oil, etc. They are widely used in foodstuffs like annatto, paprika, and saffron. At the time of food processing and storage, carotenoids are very sensitive to redox reactions and isomerization because of their electron-rich nature and highly unsaturated chemical structure (Rodriguez-Amaya 2015).

Table 8.4 Categories of caramel

Type	Name	Description
I	Plain caramel	Carbohydrates are heated with or without acids/alkalis or their salts
II	Caustic sulfite caramel	Carbohydrates are heated with or without acids/alkalis or their salts in the presence of sulfite compounds
III	Ammonia processed caramel	Food grade carbohydrates are heated with or without acids/alkalis or their salts in the presence of ammonium compounds. Sulfites avoided in this process
IV	Ammonia sulfite caramel	Carbohydrates are heated with or without acids/alkalis or their salts by using both sulfite and ammonium compounds

8.5.2.3 Annatto

It is a pigment extracted from "the pericarp of the seed of *Bixa Orellana*." It is a water-soluble compound that is yellow-orange in color. Diapocarotenoids are the basic color components in annatto. Some other degrading pigments are also available like bixin, transbixin, and norbixin. The methyl ester of a dibasic fatty acid forms the bixin pigment, which leads to the formation of hydrolyzed water-soluble norbixin on treatment with alkalis. Annatto is available in two different types: bixin (oil extract) and norbixin (powdered form).

8.5.2.4 Annatto Butter Color

An extract of annatto in solution, suspension, or oil is prepared by extracting the outer covering of seeds along with vegetable oils. It possess little or no toxicity when compared to synthetic colorants. It is used in several different ways in a number of food industries particularly in the dairy industry as it can impart color to butter and other related dairy products.

8.5.2.5 Annatto Cheese Color

The outer coating of the seed is extracted with aqueous alkali which leads to the production of water-soluble annatto color. The color of the solution in 0.1N NaOH or KOH measured in a 1-cm cell at a dilution of 1:1000 (m/v) shall be yellow units (Min.)—5.0; red units (Min.)—0.4. The color of annatto cheese is bright yellow to slight orange.

8.5.2.6 Paprika Oleoresin

It is a water-soluble extract from sweet red pepper. It is orange-red in color. Xanthophylls like capsanthin, capsorubin, and beta-carotene are the basic coloring compounds. This foodstuff is limited because of the flavoring and spicy pungency components. Raw, unrefined palm oil, which is used as a colorant for margarine, contains 0.05–0.2% carotenoids as the main constituents.

8.5.2.7 Saffron

It is a water-soluble pigment. It contains some components like crocin, digentiobioside of apocarotenic acid, and beta-carotene as flavoring compounds. The yellow color of saffron is responsible for attractive colors of beverages, cakes, and other bakery products but its high cost is a restraint in the market (Jeszka and Jadwiga 2006).

8.5.2.8 Carrot Extracts

The different types of market products based on carrots, like carrot extract, oils derived from carrot, and plant-related extracts are the colorants. The main components are alpha- and beta-carotene. Purified crystalline products are constituted up by 20% alpha-carotene and 80% beta-carotene. Fat-based products such as dispersion of microcrystals in oil are the basic use of these carrot extracts. The carotenoids when combined with surface active agents form micro-emulsions, which are important to color foods that contain higher amounts of water.

8.5.2.9 Chlorophyll

It is the green pigment extracted from plants and widely used as a coloring agent for food items. In higher plants, chloroplast is the source of this green pigment, while blue-green algae and bacteria are other sources of chlorophyll. It is located in the intracellular lamellae. It can absorb light between 400 and 500 nm (Rehman et al. 2020). Polypeptides, phospholipids, and tocopherols form complex with the chlorophyll in the plant tissues and they are held within the hydrophobic membrane. A typical leaf contains about 2.5 mg/g of total chlorophyll, 0.3 mg/g of xanthophyll, and 0.15 mg/g of carotene. The unripe state of fruit also contains chlorophyll, which gradually disappears during ripening because yellow and red carotenoids overexpose.

Chlorophylls are derivatives of dihydroporphyrin chelated with a centrally located Mg^{2+} ion. Copper and zinc ions help in retaining the green color. Preparation of food colorant from chlorophyll is mainly obtained from alfalfa and nettles. Brown seaweeds are single-cell phytoplankton and are a good source of chlorophyll. They mainly contain chlorophyll C, which is more stable than the rest of chlorophyll. Some solvents like acetone, methanol, ethanol, and chlorinated solvents are used to extract chlorophyll (Vermeulen et al. 2015). The yield of extraction is 20%, which include pheophytins and other degradation products along with chlorophylls.

Chlorophyll is of two types, i.e., water soluble and oil soluble in nature. They are stable in light, heat, and are even stable with acids and alkalis (Rehman et al. 2020). They find application as food colorants in canned products, confectionery, soups, and dairy products.

8.5.2.10 Anthocyanin

Anthocyanin being a dominant pigment in a number of fruits and flowers is responsible for providing the red, violet, or blue color. Some reactions affect the color of anthocyanin like the substitution of the hydroxyl and methoxyl groups. The increase in the concentration of the hydroxyl group provides a deep blue color whereas the methoxyl group concentration increases the redness of the food product. Examples of naturally occurring anthocyanidins include Pelargonidin (orange color), Cyaniding, Peonidin (orange-red color), Delphinidin, Petunidin, and Malvidin (bluish-red color). Anthocyanin-rich fruits include blackberry, bilberry, blackcurrant, chokeberry, cranberry, strawberry, and red grapes. The stability of anthocyanin as a colorant is poor. Anthocyanins have low stability because these pigments are very sensitive to light, heat, oxygen, and pH which limits their use as food colorants (Dangles and Fenger 2018). These are generally used at a low pH, nearly in the range of 4.0.

8.5.2.11 Beetroot Red

It is available in liquid and powder form. Beetroot contains some synthetic colorants like betanin, isobetanin, prebetanin, and smaller quantities of vulgaxanthin I and II. These are used in dairy products because they have low acid concentration. Redness can be achieved by using red beet, which is also considered as a replacement for nitrite. Antimicrobial activity is present in nitrite, but it is absent in beet root.

8.5.2.12 Riboflavin

Riboflavin is a bitter in taste, orange-yellow crystalline powder which shows green-yellow fluorescence in solutions. It is highly soluble in hydrophilic solvents like water and ethanol. It is highly stable in acidic conditions when exposed to light. Riboflavin is an essential micronutrient in the human diet, acting as precursor of flavine adenine dinucleotide and flavin mononucleotide, which in turn functions as hydrogen carrier in biological redox processes (FAO 2021). Reduction leads to the formation of a colorless leuco form, whose color gets regenerated when comes in contact with air. The physiologically active form of vitamin B2 is a riboflavin-5′-phosphate sodium salt. It is highly soluble in water as compared to riboflavin, but highly unstable in the presence of light. Both of these forms can be used as coloring agents and food additives in dressings and cheese production. The market for natural pigments is rapidly expanding, necessitating more sustainable production. Vegetable waste is the ray of hope for natural colorants for the recovery of natural pigments that are in the food, pharmaceutical, and cosmeceutical sectors (Sharma et al. 2021).

8.5.2.13 Synthetic Colorants

Organic synthetic colorants are being used in a number of industrial processes. There are more than 80 synthetic colorants present to date. These are generally used as food additives without any proper guidelines; as a result, this may lead to many health problems, intoxication, and death. Synthetic compounds are stable, easy to produce, less expensive, and have better coloring properties as compared to natural colorants. Strict rules are subjected to the use of synthetic colorants. As synthetic colorants are advanced in color supply, strength, stability, and ease of application, these are highly used. They have low market prices and provide a great spectrum of colors.

There is a trend of using fewer synthetic colorants as it increases the toxicity of food products. Many countries limit the type, use, and amount of colorants in food products. The triarylmethane group constitutes the triarylmethane dyes, and the main representative of such dyes includes bright blue, fast green, patent blue, and bright black. These compounds are generally used in dairy products like yogurt, ice cream, drinks, and sweets (Unsal et al. 2015). Indigo dye is a synthetic colorant that is chemically obtained and used in gelatin, ice cream, jelly beans, bubble gum, and candies (Deroco et al. 2018). A few examples of synthetic colorants are framed in Table 8.5.

8.5.2.14 Preparation of Food Color Solutions

The homogeneous solution can be prepared by dissolving the powder or granules in hydrophilic solvents like propylene glycol or glycerine or water, as recommended by

Table 8.5 Examples of synthetic colorants

S. no.	Synthetic colorants	Uses
1	Allura red	Bitter soda, bitter wine, and other nonalcoholic flavored drinks, flavored milk, flavored cheese
2	Amaranth	Wine and spirit drinks
3	Azorubine	Americano, bitters and wine, nonalcoholic flavored drinks, fine bakery wares, flavored processed cheese, flavored milk products, edible cheese rind
4	Brilliant blue	Mushy and canned garden peas
5	Erythrosine	Candied cherries, dietary supplements
6	Fast green	Dietary products
7	Green S	Jam, jelly, marmalades, low caloric products, processed cheese
8	Tartrazine	Americano, bitter soda, and wine

the manufacturer. It is important to select a suitable organic solvent which depends on the compatibility of the solution with the target pigment, toxicity, cost, and recovery (Souza Mesquita et al. 2021). For the stability of food colorants, specific diluents and matrices like lecithin; emulsifiers like polysorbates, diglycerides, glycerine; esters of sucrose with fatty acids; sorbitan; etc. are being used. The limiting factors of natural colorants are the color stability in the watery environment and the presence of light.

8.5.2.15 Dairy Products Using Natural Colorants

Dairy-based companies are producing beverages using flavored and anthocyanin-rich natural juices to color milk, whey-based and fermented products. Hydrophilic extracts, carmine, and beet juices are used as colorants for yogurt. These are generally resistant to microbial contamination. These colorants are generally stable at low pH and are used in dairy products. The most commonly used red colorants in yogurt are acid-proof cochineal extract and water-soluble carmine. These are heat stable at low pH, i.e., 2–7, and produce a bright color. Various examples of colorants in dairy-based food products are shown in Table 8.6.

8.5.3 Antioxidants

The molecule which inhibits the oxidation of another molecule, especially free radicals, is termed as an antioxidant. Antioxidants are important for protection from oxidative damage and maintaining the shelf life of the dairy products. Nowadays, consumers prefer natural antioxidants as synthetic antioxidants have carcinogenic effects and are toxic in nature (Abdel-Hameed et al. 2014). The formation of free radicals in food products can be controlled by natural antioxidants from plant sources. These antioxidants can avoid side effects like liver damage and carcinogenesis caused by synthetic antioxidants (Meenakshi et al. 2009). On the basis of origin they are divided into natural and synthetic antioxidants whereas on the basis of microbial production these can be classified into primary and secondary antioxidants.

Table 8.6 Various colorants used in dairy food products

S. no.	Dairy food products	Colorants used
1	Butter	Carrot extract
2	Cheese	Carotene, bixin, norbixin
3	Sausages	Curcumin, turmeric, riboflavin, carminic acid, caramel colors
4	Malt bread	Caramel colors
5	Yogurt	Carmine, beetroot color

8.5.3.1 Primary and Secondary Antioxidants

Primary antioxidants are produced by microbes and have an essential role in free radial termination, whereas secondary antioxidants are not directly formed by microbes. Carotenoids, retinol, and alpha-tocopherol are the leading lipid-soluble antioxidants present in milk, whereas ascorbic acid is the water-soluble antioxidant (Song et al. 2013). Primary antioxidants work by donating an H-atom or by a single electron transfer mechanism, and secondary antioxidants function by retarding chain initiation (Zeb 2020).

8.5.3.2 Natural and Synthetic Antioxidants

Antioxidants are broadly classified into natural and synthetic antioxidants based on their origin. Natural antioxidants are produced from natural sources and used in casein, cream, cheese, milk, etc. Butylated hydroxyanisole, propyl gallate, and butylated hydroxytoluene are examples of synthetic antioxidants which are not directly produced by microbes. These help in controlling lipid oxidation and off-flavor development by free radicals in food (Zeb 2020). A high concentration of synthetic antioxidants is harmful and carcinogenic, resulting in the high demand preference for natural antioxidants in the food industry (Kumar et al. 2019). A few examples of natural and synthetic antioxidants are shown in Table 8.7.

There is always a high demand for foods such as fruits, vegetables, grains, and dairy products. This high demand for antioxidants is completed by increasing product retention periods so that fresh and nutrient-rich food is available to consumers. Antioxidants not only extend food shelf life but also protect against oxidative stress and inflammation. The entire dairy products need antioxidants for better life. Nowadays, both primary and secondary antioxidants are in industrial use.

8.5.3.3 Antioxidants in Milk

The oxidation of milk is mainly counterbalanced by antioxidants like whey protein, casein, vitamin C, vitamin E, enzymes like superoxide dismutase, catalase, glutathione peroxidase, etc. Casein is the major protein of bovine which inhibits lipoxygenase-catalyzed lipid autooxidation. The primary structure of the casein

Table 8.7 Examples of natural and synthetic antioxidants

Examples of antioxidants	
Natural antioxidants	Synthetic antioxidants
Whey protein	Butylated hydroxyanisole
Casein	Propyl gallate
Vitamins (C, E, D)	Butylated hydroxytoluene

molecule acts as a scavenger for free radicals (Suetsuna et al. 2000). In sheep's milk, protein has been considered as the main antioxidant (Fardet and Rock 2017). It acquires antioxidant capacity because of casein, beta-lactoglobulin, and alpha-lactoalbumin (Fardet and Rock 2017). Vitamins A, C, and E, beta-carotene, and coenzyme-Q are natural antioxidants present in human milk. They help in neutralizing the effect of reactive oxygen and nitrogen species in milk (Khan et al. 2019). Lipophilic antioxidants help in preventing oxidative stress by activating the defense system in milk (Lindmark-Månsson and Akesson 2000). They have high level of thermal stability and remain active in dairy products.

8.5.3.4 Antioxidants in Processed Cheese

Food products made up of cheese and unfermented dairy ingredients are mixed with emulsifiers. Vegetable oils, salt, and sugar are some additional ingredients added to increase the color, texture, and flavor of cheese. These are used in many dishes as processed cheese does not separate when melted. It is a good source of protein, fat, minerals, and vitamins in the diet that need to be preserved by antioxidants (Abdel-Hameed et al. 2014). The antioxidant activity of peptide was checked in Mexican cheese after 6 months and characterized by HPLC. It was observed that peptides with antioxidant activity were produced during the ripening period of 6 months and the cheese showed 98% radical scavenging activity with 2,2-diphenyl-1-picrylhydrazyl (DPPH) (Haug et al. 1987).

8.5.3.5 Antioxidants in Cream

Cream is the fatty part of nonhomogenized milk that floats on the top layer of milk. Generally, cream contains 10–65% of milk fat used in various products like coffee, pies, soups, etc. Cream is classified into several types based on its milk fat content, including clotted cream (55% milk fat), heavy cream (>35% milk fat), whipping cream (30–36% milk fat), light cream (18–25% milk fat), sour cream (18% milk fat), and so on. Lipophilic antioxidants like phospholipid, coenzyme Q10, CLA (conjugated linoleic acid), etc. play a key role. The high use of emulsifiers increases the demand for lipophilic antioxidants ten times (Niki 2014).

8.5.3.6 Antioxidants in Yogurt

Fermentation of milk with the help of bacteria leads to the formation of yogurt. It is also spelled yoghurt and is made up of 81% water, 9% protein, 5% fat, and 4% carbohydrates, including sugars. The bacteria used to make yogurt are also known as "yogurt cultures." Ferric Reducing Antioxidant Power (FRAP) and 2,2-diphenyl-1-picryl-hydrazyl-hydrate (DPPH) assay were used to check the antioxidant activity of yogurt mixed with carrot, pumpkin, broccoli, and red sweet pepper with 10%

Table 8.8 Various antioxidants in dairy products

S. no.	Dairy-based products	Antioxidants
1	Milk	Whey protein, casein, vitamin C, E, glutathione peroxidase, superoxides dismutase, carotenoids, etc.
2	Processed cheese	2,2-diphenyl-1-picrylhydrazyl, peptide, carotenoids
3	Cream	Phospholipid, coenzyme Q10, CLA (conjugated linoleic acid), carotene
4	Yogurt	2,2-diphenyl-1-picrylhydrazyl, alpha-casein, beta-carotene, vitamin C, and E

concentration of antioxidant and observed for 14 days. High antioxidant activity was seen in yogurt with broccoli and red pepper. The antioxidant activity was decreased by the 14th day (Najgebauer-Lejko et al. 2014). Studies have also been conducted to check the antioxidant property of yogurt with vitamin C, vitamin E, beta-carotene, ascorbic acid, and even alpha-casein. A few examples of antioxidants in dairy products are formulated in Table 8.8.

8.5.4 *Flavoring Agents*

Food becomes cherishing because of its flavor. Flavoring agents are the complex mixture of large amounts of low-molecular-weight volatile compounds (<300 g/mol) capable of stimulating the odor receptors in the nose (Parker et al. 2014). Natural flavors are formed by the biosynthesis of aromatic chemicals during the normal metabolic processes in plants and animals. The odor and taste of these aromatic components represent the intrinsic flavor of food. Flavoring is man-made. It is prepared by adding natural or synthetic aromatic substances that may or may not be present in nature. The aim is to increase the acceptability of the end product through the stimulation of the nose and the palate, by modifying a flavor that is already present or masking some undesirable flavor.

8.5.4.1 Functions of Flavoring Agents

Flavoring agents have three major functions:

Economic

Some nutritional foods may have an undesirable taste and smell, e.g., vitamins and soya. Flavors are used to modify the taste, so that nutritional food can become more desirable and this ultimately helps in increasing the economic value of the food product (Sinki and Gordon 2001). Flavoring agents can recompense the flavor losses that occur during the processing and storage of food items in order to maintain their freshness and shelf life.

Physiological

According to the taste and fat digestion study, there are some indications that metabolic response is affected by the alteration in taste of a fatty meal. Research studies on "taste and intestinal absorption of glucose" show that "oral stimulation affects intestinal absorption." These create a direct link between a physiological condition and the flavoring agent.

Psychological

Flavoring provides the sensory pleasure. Proper flavor selection can be greatly assisted by psychological analysis of pleasure. The concept of flavoring also increases the variety of food in the market.

These three functions, i.e., economic, physiological, and psychological have different degrees of application. For example, some applications are mainly for pleasure, which comes under psychological function, or increasing the market value by adding flavor to the milk is totally an economical concept. All these compounds need a raw material for flavor formation, depending upon the requirements of the flavor.

8.5.4.2 Raw Materials for Flavoring Agents

Flavoring compounds require basic raw materials for their formation. This raw material can be of natural or synthetic origin. The natural raw materials are further divided into botanical and animal raw materials. Examples of botanical raw materials are fruits, vegetable juice, herbs, spices, nuts, etc. On the other hand, examples of animal raw materials are plasma dripping, seafood by-products, enzyme-modified cheese, meat extract, etc. whereas benzaldehyde and cinnamic alcohol are examples of synthetic raw materials used for the production of the flavoring material (Mani-Lopez et al. 2012). Generally Recognized as Safe (GRAS) status and certification are needed for the utilization of flavor in consumer products, which are further approved for use in food and cosmetics industries (Smith et al. 2005b). Maillard reaction is the basic mechanism performed for the production of flavoring compounds. It occurs when amino groups interact with the sugars. For example, milk contains lactose sugar, which participates in numerous Maillard reactions (Hwang et al. 2011). A brown nitrogenous compound, or melanoidin, along with other polymers is formed as an end product of Maillard reaction which has a specific flavor (Bastos et al. 2012).

8.5.4.3 Classification of Flavoring Agents

Based on the source of the material, flavoring agents can be divided into three broad categories:

Natural Flavoring Agents

Natural flavoring agents are obtained from plant and animal sources. These can be naturally made or produced in a lab by chemical reaction. The chemical formulae of the parent and new compound are the same in natural agents. They need to be isolated via distillation, extraction, etc. These are the processes where one particular substance is isolated from the rest of the other substances based on their physical and chemical properties. For example, vanillin, the main component of vanilla beans, can be produced as a natural or a nature-identical flavoring substance. These substances are made in the laboratory, but their chemical structure resembles the structure present in natural products. A substance can only be named as nature-identical if it naturally occurs in plant or animal raw materials. Different processes are applied for the extraction of natural flavors:

Distillation: It is a process where the plant and animal source materials are brought to a boiling point and the steam is collected via cooling.

Extraction: It is a process in which the compounds are mixed with a dissolving agent like alcohol or supercritical carbon dioxide, etc. for the extraction of flavoring compounds.

Biotechnological production: It is used in the case of source material where the simple extraction is difficult to perform and the material to be extracted is in very low quantity, or the material is highly complex for the extraction. It is an expensive process to conduct.

Artificial/Synthetic Flavoring Agents

Artificial flavoring agents are produced by chemical methods. These agents have different chemical formulae as compared to their substrate formulae. Chemical methods or synthetic methods are used to isolate these flavoring agents. These are obtained from different source material because the methods of production are different for every other agent. Ethyl vanillin, well-known artificial flavoring agent, is a substance that smells and tastes like vanillin.

Other Flavoring Techniques

(a) **Essential oil flavoring**: It involves a complex mixture with defined composition like natural raw materials. Vegetable extracts, spices, and herb extracts are examples of these substances. These are obtained from plant, animal, and microbial sources by using physical and biotechnological production processes. These are the main components of citrus, spice, and mint flavorings. It is helpful in relishing the smell and taste of the components. These categories also include oils obtained from clove and eucalyptus.

(b) **Thermal process flavoring:** This technique is used in industrial processes by controlling several parameters like heat, temperature, pH, etc. which leads to the development of intense flavors during these processes. For example: in the case of roasting meat or baking bread, the thermal process leads to the formation of these flavors. Amino acids and reducing sugars are the essential basic materials.

(c) **Smoked flavoring:** It is one of the ancient methods for the production of seasonal food and preserving food. It not only is used for preserving but also adds a unique smoked flavor to foods such as fish, meat products, sauces, and chips. Burning of hardwood at a temperature of nearly 100 °C is used to produce the fresh smoke. Charcoal is the basic product and this is mostly formed in the absence of air. Efforts have been made to reduce addiction to nicotine, which leads to the risk of different physical health problems. It also simulates the physical action of conventional cigarettes, but without using any of the nicotine products (Litt et al. 2016).

8.5.4.4 Applications of Flavoring Agents

A particular flavor can be added to give an additional taste and flavor to the food. Flavoring agents can be exploited for their applications in different industries as shown in Fig. 8.4.

Fig. 8.4 Application of flavors

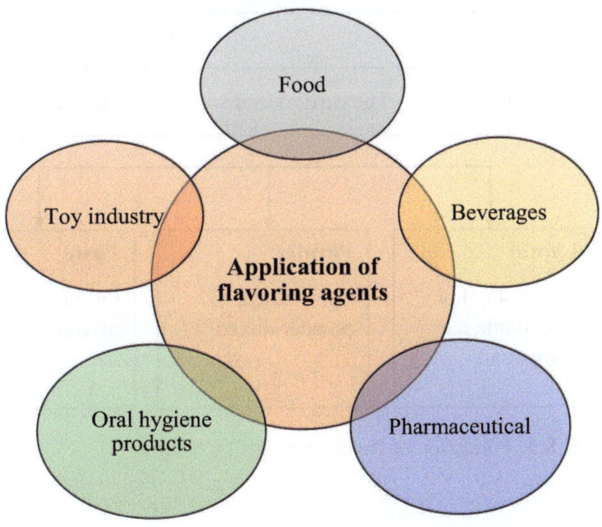

On the basis of applications of flavoring agents, food can be categorized as follows:

(a) **Flavor-dependent food** includes the food items which cannot exist without the application of flavors. For example, hard-boiled candy, non-juice drinks, gelatin desserts, powdered milk, bread, and cakes.
(b) **Flavor-independent food** is a market product that does not use flavor or is not dependent on flavor for its marketing. For example, milk, orange juice, and butter.

8.5.4.5 Flavoring Agent Formation

Flavor helps in providing the artificial taste and smell. Some of these flavors are diluted with carriers and solvents. These are of basically four types, i.e., liquid, powder, paste, and emulsion (Fig. 8.5). These liquids are made from water, alcohol, or oil-soluble compounds. Alcohol, propylene, glycol, triacetin, benzyl alcohol, glycerin, syrup, etc. are the solvents and carriers for liquid-type flavoring agent. The higher molecular weight of different mono- and disaccharides increases viscosity, which improves beverage taste (Mussatto and Mancilha 2007). Another is a powder-type flavoring agent that can be spray-dried, absorbed, or powder mixed. These use gum acacia, starch hydrolysates, selective hydrocolloids, and simple carbohydrates as solvents and carriers. Paste and emulsion can form an oil-in-water type of solution (Tandjawa et al. 2017). The solvent and carrier of paste and emulsion are the same as liquid, and powder-type flavoring agents.

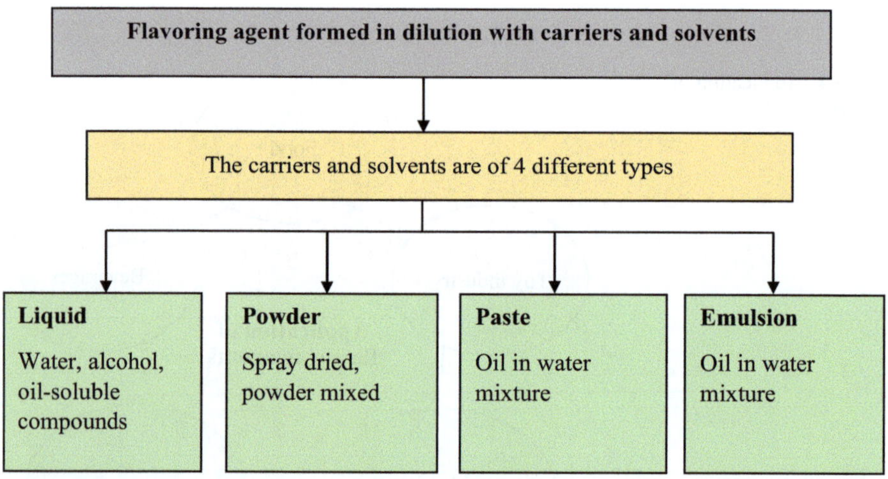

Fig. 8.5 Formation of flavoring agents by dilution

8.5.5 Thickeners and Gelling Agents

Thickeners are generally hydrocolloids in nature. The starch-hydrocolloid blending has a specific application at industrial level as an emulsifier, gelling agent, thickener, stabilizer, and for inhibiting the formation of sugar and ice crystals (Marta et al. 2020). Conformational disorder in polymer chains is the result of thickeners in the solution. It acts by polymer-solvent interaction. Thickeners, such as modified starch and xanthan gums, are primarily used in the food industry. Different types of thickeners like cornstarch, pre-gelatinized starch, arrowroot, algin (sodium alginate), agar-agar, gelatin, acacia, pectin, etc. are used in dairy-based food (Nilsen-Nygaard et al. 2016).

Gels are intermediate between solid and liquid phases with some mechanical rigidity. They work by forming an interlinking connection through a cross-linking network between the liquid and solid phases. The major bond that plays a role in this gel formation is hydrogen bonding, which increases the industrial application of starch-based compounds (Mahmood et al. 2017). Hydrocolloid gels are also known as physical gels as they are made up of covalent cross-linking of the polymer chain. Hydrocolloids are nonstarch polysaccharides which are heterogeneous in nature (Noisuwan et al. 2009). They are made up of long-chain polymers characterized by the formation of viscous dispersions and/or gels when dispersed in water (Saha and Bhattacharya 2010). The type of starch and hydrocolloids, concentration of starch and hydrocolloids, level and type of interaction, preparation condition, etc. decide the starch-hydrocolloid mixture properties (Varela et al. 2016). The interaction of starch-hydrocolloids can alter food texture, structure, and viscosity, thereby changing the accessibility of enzymes to starch granules. A few examples of gelling agents and their applications in dairy products are discussed in Table 8.9.

Table 8.9 Various gelling agents and their applications

S. no.	Gelling agents	Gel characteristics	Applications
1	Modified starch	Thermally irreversible opaque gels formed on cooling	Dairy dessert
2	Agar	Thermoreversible gels on cooling	Bakery products
3	K and I carrageenan	Thermoreversible gels on cooling	Puddings, milkshakes, tofu
4	Low methoxyl pectin	Thermoreversible gels on cooling at acidic pH	Jams, jellies, glazes, milk-based desserts
5	Alginate	Thermoirreversible gels do not melt on heating	Restructured foods, cold prepared bakery creams

8.5.6 *Humectants*

Humectants are hydroscopic substances added to food to increase the retention time of moisture. Due to the low degree of humidity, the food tends to dry out because of the counteracting effect of the wetting agent. For the formation of candied intermediate moisture fruits, sugar is the most common agent used as humectants whereas to provide intermediate moisture to vegetables and fish, salt is used as a major agent (Saha 2020). These humectants prevent the food from drying out. The basic function of humectants is to control the level of water in the solution. For the immobilization of water in food, solutes like salt and sugar are used as humectants. The drying process removes free water and humectants from the food. These humectants are made up of hydrophilic parts that form bonds with water in food. If water is removed, it leads to lower water activity of the food product (Kaplow 1970). These are used in gelatin products and cakes (Siburian et al. 2020). Some examples of humectants used in the dairy industry are provided in Fig. 8.6.

8.5.6.1 Sorbitol

It is widely present in nature, and sorbitol syrup is used as a humectant. It has the capacity to bind to a higher number of water molecules (Duan et al. 2019). It is generally used in bakery products, mayonnaise, creams, sauces, etc.

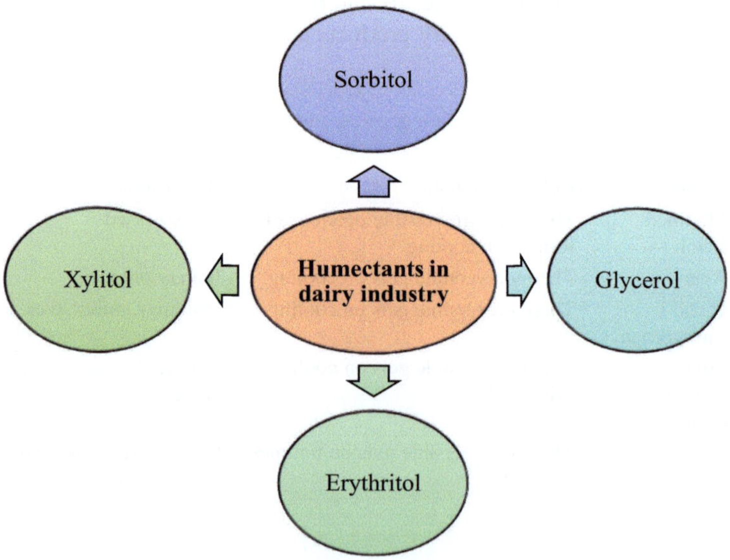

Fig. 8.6 Humectants in dairy industry

8.5.6.2 Glycerol

The hydrolysis of fat leads to the formation of glycerol. It can be readily obtained from animal and vegetable fats. It is generally liquid at room temperature. The basic function of glycerol as a humectant is to lock the food moisture and avoid the formation of mold or bacterial growth. It also helps in maintaining the texture of the food. It provides a burning sensation in the throat as a form of taste, which limits its quantities. But still, it is widely used in the cake and cream industry.

8.5.6.3 Xylitol

Xylitol and sucrose have an equivalent sweetness; hence, xylitol is primarily used as a sugar substitute in many food products (Ahuja et al. 2019). It is wood sugar that is collected from xylan-rich hemicelluloses of a tree or a plant. It generally acts as a sweetener but for nonfermentable bulk. Because of the high demand for xylitol in food, nutraceuticals, pharma industry, and beverages, it is among the top 12 global bio-products (Vishal et al. 2019). It is also used as an energy source in many intravenous products. It is also used in dental products like toothpaste, mouthwash, creams, etc. and it is well used in dairy products (Ahuja et al. 2020).

8.5.6.4 Erythritol

It belongs to the category of polyols generally used as a sweetener and a humectant. Sometimes it also acts as a flavor enhancer, stabilizer, thickener, or sequestrant in the food industry. The main products of erythritol are desserts, food supplements, liqueurs, etc.

8.5.7 Emulsifying Agents

An additive that allows immiscible liquids like water and oil to form a stable mixture is known as an emulsifier. When microscopic droplets of one immiscible liquid mix with another immiscible liquid, it leads to the formation of a heterogeneous emulsion (McClements 2015). An emulsion is a mixture of immiscible liquid. The boundary between two nonhomogenous mixtures is known as interface. In the mixture, the dispersed phase droplets tend to coalescence and agglomerate, which ultimately leads to separation. Starch and soy protein are the solid and colloidal particles which get absorbed at the interface and form the Pickering emulsion (Yang et al. 2017). The emulsifier contains both hydrophilic and hydrophobic parts. The hydrophilic part makes the bond with water or polar compounds, whereas the hydrophobic part gets dissolved with oil or nonpolar compounds. The hydrophilic part makes the head and tail, which is made by the hydrophobic part of the micelle (Berton-Carabin et al. 2018).

Glycerol, sorbitol, sucrose, propylene glycol, or polyglycerol forms the hydrophilic part of commercial food emulsifiers whereas fatty acids derived from fats and some oils like soyabean oil, rapeseed oil, coconut oil, etc. form the lipophilic part of the emulsifier. The dispersion of fat droplets in water leads to the formation of an oil-in-water emulsion, e.g., mayonnaise and milk. When water is in the dispersion phase and fat acts as a continuous phase, such as in butter and vinaigrettes, the result is a water-in-oil emulsion (Leal-Calderon et al. 2007).

8.5.7.1 Stability of an Emulsion

Stability of an emulsion is best described by its droplet size, viscosity, and density.

Droplet size: emulsion stability is increased by a smaller droplet size. Homogenization is the process that results in even and smaller droplet sizes.

Viscosity: Emulsion stability is facilitated by a higher viscosity. Thickening agents, or hydrocolloids, are responsible for increasing the viscosity of the water phase (Krstonosic et al. 2021).

Specific density of phases: An emulsion is more stable if the density difference between the two phases is less.

Food emulsifiers have versatile applications in the food industry. The functions of emulsions in dairy-based products are given in Table 8.10.

8.5.7.2 Properties of Emulsifier

An emulsifier should be readily absorbed into the interface, should have proper hydrophilic and lipophilic parts so that a stable emulsion can be formed, should be stable, should not have any odor, color, taste, etc., and should be nontoxic.

8.5.8 Antifoaming Agents

An antifoaming agent is also known as a deforming agent. It is a chemical that hinders or reduces the formation of foam in industrial processed liquids. Detergents,

Table 8.10 Function of emulsions

S. no.	Functions of emulsions	Dairy-based foods
1	Improved solubility	Instant drinks
2	Plasticizers	Cake icing
3	Promotion of dryness	Ice cream
4	Oil in water emulsion	Mayonnaise
5	Stabilization and avoiding the formation of an oil layer	Coffee whitener
6	Improve texture and taste	Dairy desserts

Table 8.11 Some common antifoaming agents

S. no.	Antifoaming agents	Dairy-based function
1	Polyoxyethylene sorbitan tristearate (polysorbate 65)	Antifoaming agents in ice cream, frozen desserts, and cakes
2	Sodium, potassium salts of fatty acids	Yeast activity promoter and antifoaming agent in beet sugar
3	Acetic acid esters of fatty acids	Antifoaming agents in jams, marmalades, and acetems
4	Sorbitan esters	Antifoaming agent in production of beet sugar and boiled sweets
5	Dimethylpolysiloxane	Antifoaming agent in skim milk powder and wine production
6	Propan-1,2-diol	Anticaking agent, soft drink antifoaming agents

pharmaceutical, food, and processing industries require detergents as an antifoaming agent. This foam is generally produced because of protein and fat in the food. To completely suppress the foam, 10 ppm of antifoam is generally used. These compounds are used in meat processing, dairy products, vegetable oil, alcoholic and nonalcoholic beverages, soups, starches, pickles, etc. A few examples of antifoaming agents with their dairy-based functions are provided in Table 8.11.

8.5.9 Stabilizers

Food additives, which help in preserving the structure of food products, are known as stabilizers. It prevents the formation of large ice crystals in ice cream. It helps in maintaining the physico-chemical state of the food product. It provides uniformity to the product and increases the water holding capacity of the food. The amount of stabilizers to be used varies with the properties of the food and percentage of solid content in the food. Stabilizers are water-soluble compounds majorly used in dairy-based foods like ice cream, processed milk, pasteurized milk, etc. On the other hand, the addition of excessive amount of stabilizers results in soggy or heavy body and high resistance to melting point in the final product.

Ideal characteristics of a stabilizer are shown in Fig. 8.7.

8.5.9.1 Different Stabilizers Used in Dairy Industry

Nowadays, the market contains a wide range of hydrocolloids derived from plants and seaweed, as well as those produced by various microbes such as bacteria, fungi, and molds. Hydrocolloids are made up of sugar backbones that contain protruding substituent such as esters, sulfates, or added sugars. Available hydrocolloids for food applications are either neutral or negatively charged. A few examples of stabilizers used in the food market are depicted in Fig. 8.8.

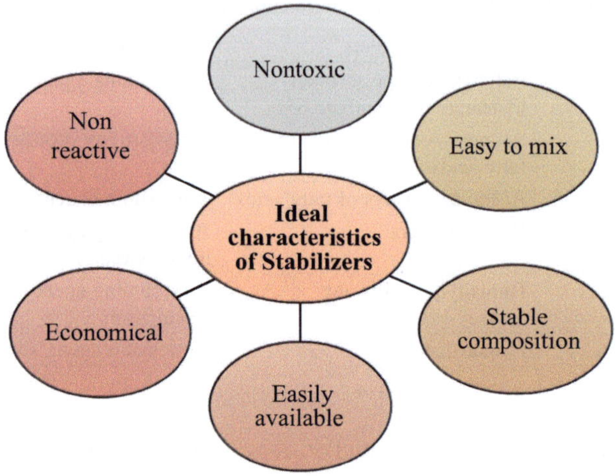

Fig. 8.7 Ideal characteristics of stabilizers

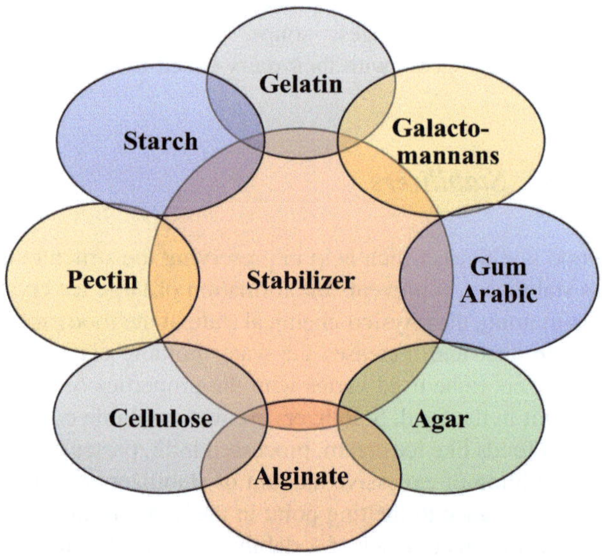

Fig. 8.8 Different stabilizers used in dairy industry

Gelatin

Animal connective tissues, mainly collagen, provide the proteinaceous material, i.e., gelatin. The raw materials for gelatin production are provided by beef, pork, poultry, and fish. It provides a low viscosity solution when hydrated in hot or warm water, which has good foaming and whipping properties (Siburia et al. 2020). Clear and elastic gels are formed by the network of polypeptide chains in gelatin. It is a good stabilizer but does not prevent heat shock. One more advantage of this stabilizer is that it requires a long aging period. It does not cause wheying off and has good dispersal activity. Gelatin is utilized in food industries with various applications like gelling agents, thickeners, emulsifiers, foam makers, stabilizers, edible films, etc. It is also widely used for the manufacturing of soft and hard capsules in the pharmaceutical industry (Park et al. 2008).

Galactomannans

Galactomannans belong to the category of naturally occurring biocompatible and biodegradable molecules. These are nonionic polysaccharides made up of mannose and galactose residues (Yadav and Maiti 2020). These are used in different dairy products such as cheese, processed meats, bakery products, and frozen desserts. The most widely used galactomannans are locust gum and guai gum. They are differentiated based on solubility in cold water as well as gelling capability because of their difference in distribution side and substitution degree (Yadav and Maiti 2020).

Gum Arabic

It is also known as Acacia gum, naturally obtained from acacia trees. The main bonding in gum Arabic is of polysaccharide from (Wu et al. 2013a; Wang et al. 2015; Amin et al. 2013) (1–6)-linked β-D-galactopyranosyl units along with (1–6)-linked β-D-glucopyranosyl uronic acid units. It has relatively low viscosity and high water solubility compared to other exudate gums (Moneim and Sulieman 2018).

Agar

Agar is a linear polysaccharide obtained from red-purple algae of the Rhodophyceae family. It is made up of disaccharide units of (1–3)-linked β-D-galactose and (1–4)-linked 3,6-anhydrous-a-L-galactose residues (Mostafavi and Zaeim 2020). As a result of its characteristics like gelling and stabilizing properties, it is widely used as a valuable material in various applications (Park et al. 2020). Syneresis is the strong gel formed by agar. It can form reversible gels simply by cooling and providing heat (Motafavi and Zaeim 2020). The addition of salts can improve the high melting point of agar.

Alginates

These are produced from brown seaweeds, i.e., Phaeophyceae. In the market, the salt derivative of alginate is known as algin. These are unbranched copolymers of (1–4)-linked β-D-mannuronic acid (M) and α-L-guluronic acid (G) residues. Alginate dissolves in cold water and gels in the presence of calcium and acid. Acidified milk is stabilized by using casein (Li and Zhong 2020).

Cellulose and Its Derivatives

When natural cellulose is treated with hydrochloric acid, it leads to the production of microcrystalline cellulose, i.e., MCC which form the crystallite aggregates as an end product. Carboxy-methyl cellulose (CMC) is an anionic, water-soluble polymer that leads to the formation of various viscous solutions. Treating cellulose with alkali and then with monochloroacetic acid forms CMC. It is a carboxy-methylated derivative of cellulose, generally present as a sodium salt (Du et al. 2008). Sometimes it is also known as "guar gum." Sodium carboxy-methyl cellulose is a chemically modified natural gum that is generally linear, water-soluble, long-chain, and anionic. The degree of carboxy-methyl substitution of CMC as a food additive is around 0.6–0.95, which makes it negatively charged in acidic conditions. CMC can be adsorbed onto the casein micelle surface (pH 3.0–5.2) to provide electrostatic repulsions (Wu et al. 2013a, b). This blend is a tasteless, water-soluble, free-flowing, cream-colored, and odorless powder.

Pectin

Some food industry by-products, such as apple pulp, citrus peels, and sugar beet pulp, aid in the identification of commercial pectin. Pectin is a complex polysaccharide present in plant cell wall. Pectin from citrus sources is rich in galacturonic acid (GalA), and this GalA is present as a linear structure with the main structure of the homogalacturonan (HG) domain, while the branched rhamnogalacturonan domains (RG-I and RG-II) constitute only a small part and are interspersed in HG (Zhang et al. 2018). Aggregation and sedimentation of casein in acidified milk drinks is prevented by using pectin (Doesburg and De Vos 1959). Galacturonans and rhamnogalacturonans are the two main families of pectin which are being used as additives. The degree of esterification depends on the category of pectin. These are divided into two categories: low methyl (LM) pectin that contains less than 50% methyl esters and high methyl (HM) pectin with more than 50% methyl esters.

Starch and Modified Starches

Starch is a polysaccharide which consists of a large number of glucose units bonded together with the help of glycosidic bonds. Starch is basically composed of amylose and amylopectin, and these two polymers are packed into semi-crystalline granules which are highly hydrophilic and characterized by swelling properties (Agoda-Tandjawa et al. 2017). Starch has limited use in the food industry. Generally, starch produces weak, cohesive, and rubbery pastes on heating, and undesirable gels are produced. These properties are modified with the help of various modifications.

Some examples of modified starches are acid-treated starch, bleach starch, alkaline-treated starch, oxidized starch, monostarch phosphate, distarch glycerol, distarch phosphate esterified, etc. Water holding capacity and viscosity of starch can be increased by embedding starch granules inside the casein matrix (Azim et al. 2010). The ratio of substituent groups is kept low in modified starch. Maltodextrin is a part of starch with a low molecular weight. It is also known as dextrin or amylolytic enzyme. Pre-gelatinization, or heat treatment, of starch is the general part of the physical modification. It is sometimes used as a free thickener in cold water as well. Chemical modification (viz. acid/alkaline treated, oxidized, etc.) has been considered as mainstream of modified starch in the last century.

8.6 Disadvantages of Additive Manufacturing

Food additives are being used in beverages, desserts, jams, jellies, sauces, pickles, cosmetics, toothpaste, etc. Despite their wide use, they can cause some issues in the body. Artificial colorants can cause attention deficit disorder (ADD), inhibit the immune system, and lead to hyperactivity and allergic reactions. The overuse of colorants can also cause thyroid tumors, urticaria dermatitis, asthma, nasal congestion, nausea, eczema, liver damage, kidney damage, digestive disorders like diarrhea and colicky pains, and a wide range of cancers. Hives, itching, rashes, and swelling are some of the skin disorders caused by additives.

8.7 The Legal Aspects of Food Additives

Due to the increasing use of food additives in day-to-day life, it is very important to provide basic guidance on the maximum use of additives based on dietary pattern. Joint Expert Committee on Food Additive and Contaminants (JECFA) is an international expert scientific committee that is administered jointly by the Food and Agriculture Organization of the United Nations (FAO) and the World Health Organization (WHO) to evaluate the toxicological data on food additives and when considered to be appropriate, the committee establishes acceptable daily intakes (ADIs) on the food additive.

GRAS stands for "Generally Recognized as Safe." It is a group that qualify food products as safe to use. These food products are considered safe to use because they have a long history and have not shown any harmful effects. Obtaining approval of a substance like a food additive generally requires the safety of scientific procedures that require the same quantity and quality of scientific evidence for recognition (Smith et al. 2005a).

8.7.1 Good Manufacturing Practices (GMPs)

These are the guidelines of provision and standards which are to be used under the condition of good manufacturing practice. According to these guidelines, food additives used in the manufacturing, processing, or packaging of food should have

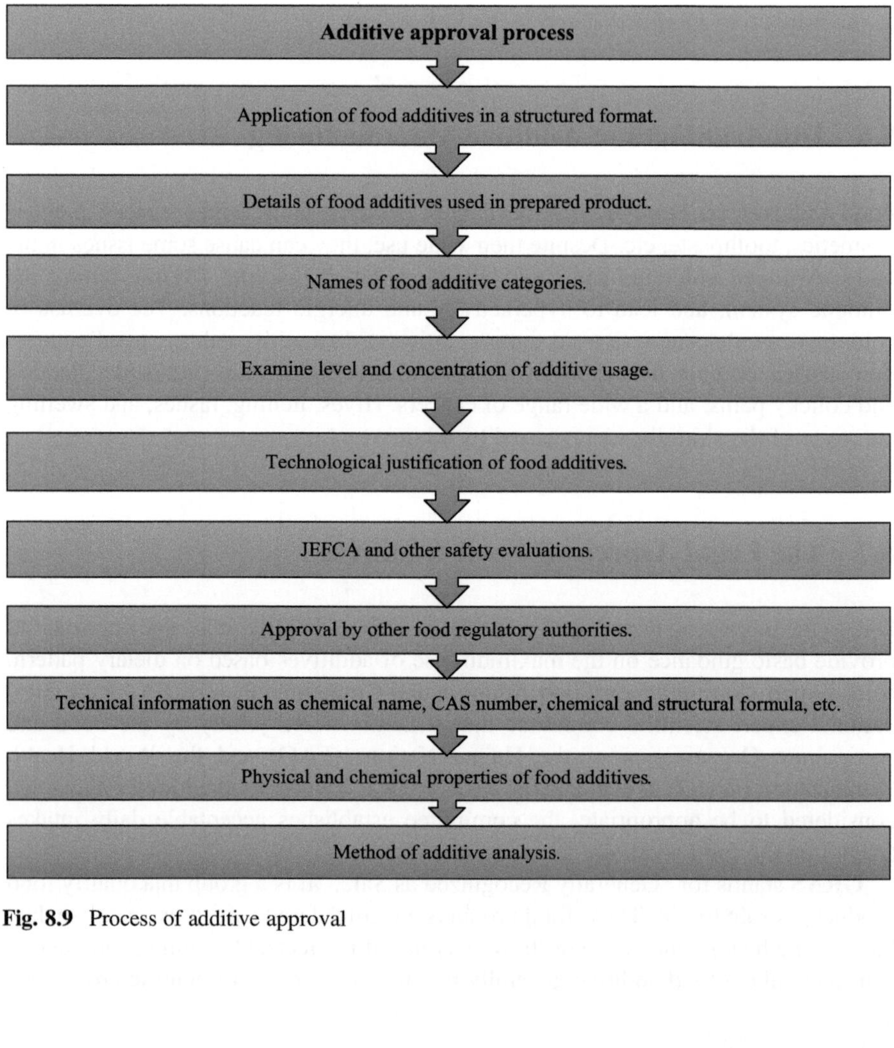

Fig. 8.9 Process of additive approval

no physical, chemical, or technical side effects. The additive should be handled in the same way as a food product is handled. The quantity of additives should be low.

In India, these guidelines are created and maintained by the FSSAI Act and rules. Some of the basic food additive approval processes in India are given below in Fig. 8.9.

8.8 Conclusion

Maintaining and increasing the shelf life of food for industrial applications by using additives is a reasonable approach. Some chemical additives which are widely used in different dairy products like milk, cream, yogurt, cheese, custard, casein, fermented soured milk, etc. are natural and others are synthetic, Not only this, additives are also widely used in different traditional dishes like ymer, villi, mursik, kumis, etc. Natural preservatives and antimicrobial agents like acetic acid, lactic acid, lysozyme, natamycin, nutella, etc. are widely used to control bacteria, molds, fungus, etc. Natural and synthetic colorants are used in fermented drinks and flavored milk. Aside from this, antioxidants reduce the oxidation of food by free radical mechanism. Flavors widely play a vital role in economic, physiological, and psychological factors. Currently their application is even extended to toy industries. Thickeners are hydrocolloids which are used in emulsifier, gelling agent, thickener, stabilizer, and for inhibiting the formation of sugar and ice crystals. Emulsifying agents, stabilizers, and antifoaming agents are used in batch and continuous fermenters at the industrial level. Despite this tremendous potential, these additives still have limitations because of their toxicity. Many agencies like ADI, GRAS, and FSSAI regulate industrial use and provide guidelines from time to time for the better use of additives. But still, there is a high need for natural substitutes as additives for long-term use.

References

Abdel-Hameed E-SS, Nagaty MA, Salman MS, Bazaid SA (2014) Phytochemicals, nutritionals and antioxidant properties of two prickly pear cactus cultivars (Opuntia ficus indica Mill.) growing in Taif KSA. Food Chem 160:31–38. https://doi.org/10.1016/j.foodchem.2014.03.060. S0308814614004518

Agoda-Tandjawa G, Le Garnec C, Boulenguer P, Gilles M, Langendorff V (2017) Rheological behavior of starch/carrageenan/milk proteins mixed systems: role of each biopolymer type and chemical characteristics. Food Hydrocoll 73:300–312. https://doi.org/10.1016/j.foodhyd.2017.07.012. S0268005X1730632X

Ahuja V, Sharma A, Rathour RK, Sharma V, Rana N, Bhatt AK (2019) In-vitro and in-silico characterization of xylose reductase from Emericella nidulans. Curr Chem Biol 13(2):159–170. https://doi.org/10.2174/2212796812666180622103906

Ahuja V, Macho M, Ewe D, Singh M, Saha S, Kumar S (2020) Biological and pharmacological potential of xylitol: a molecular insight of unique metabolism. Foods 9(11):1592. https://doi.org/10.3390/foods9111592

Amin L, Azad M, Samian A (2013) Factor influencing risk perception of food additives. J Food Agric Environ 11:66–72

Amiri S, Aghamirzaei ZM, Mostashari P, Sarbazi M, Tizchang S, Madahi H (2020) The impact of biotechnology on dairy industry. In: Microbial biotechnology in food and health, vol 33. Academic, New York, pp 53–79

Azim Z, Alexander M, Koxholt M (2010) Influence of cross-linked waxy maize starch on the aggregation behavior of casein micelles during acid-induced gelation. Food Biophys 5:227–237

Bacak A (2017) Acidity regulators, preservative, and antioxidants. Wiley Online Library. https://doi.org/10.1002/9781118906460.ch1g

Bastos DM, Monaro É, Siguemoto É, Séfora M (2012) Maillard reaction products in processed food: pros and cons. In: Food industrial processes—methods and equipment. IntechOpen, London

Benucci I, Lombardelli C, Mazzocchi C, Esti C (2022) Natural colorants from vegetable food waste: recovery, regulatory aspect, and stability—a review. Compr Rev Food Sci Food Saf 21(3):2715–2737. https://doi.org/10.1111/1541-4337.12951

Berton-Carabin CC, Sagis L, Schroën K (2018) Formation, structure, and functionality of interfacial layers in food emulsions. Annu Rev Food Sci Technol 9:551–587

Brennan CS (2005) Dietary fibre, glycaemic response, and diabetes. Mol Nutr Food 49(6):560–570

Cai W, Liu J (2014) The improvement of the risk communication mechanism in food additives under the perspective of consumers' right to know. J Food Saf 5:167–172

Carmen S (2007) Natural pigments as food colorants, chemical and functional properties of food components. https://doi.org/10.1201/9781420009286

Chen J, Han F (2009) Interpretation of highlights in the Chinese food safety law from a scientific perspective. Chin J Prev Med 43:470–474

Chen S, Wu H, Lu X, Zhong K, Xie X, Li X, Luo X, Guo L (2015) The public's risk perception on food additives and the influence factors. J Chin Inst Food Sci Technol 15:151–157

Christensen T, Mørkbak M, Jensen S, Evald J (2011) Danish consumers' perceptions of food additives and other technologies. Inst Food Resour Econ 4:23–35

Codex Alimentarius Commission (2017) General standard for food additives: codex stan 192–1995. International Food Standards

Cortez R, Luna-Vital DA, Margulis D, Gonzalez de Mejia E (2017) Natural pigments: stabilization methods of anthocyanins for food applications. Compr Rev Food Sci Food Saf 16(1):180–198

Dangles ON, Fenger JAN (2018) The chemical reactivity of anthocyanins and its consequences in food science and nutrition. Molecules 23(8):1970

Davidson PM, Branen AL (2005) Food antimicrobial an introduction. Food Sci Technol. https://doi.org/10.1201/9781420028737.ch1

Davidson M, Juneja V, Branen J (2001) Antimicrobial agents, food science and technology. https://doi.org/10.1201/9781420028737.ch1

de Souza LM, Martins M, Pisani LP, Ventura SP, de Rosso VV (2021) Insights on the use of alternative solvents and technologies to recover bio-based food pigments. Compr Rev Food Sci Food Saf 20(1):787–818

Deroco PB, Medeiros RA, Rocha-Filho RC, Fatibello-Filho O (2018) Selective and simultaneous determination of indigo carmine and allura red in candy samples at the nano-concentration range by flow injection analysis with multiple pulse amperometric detection. Food Chem 247:66–72

Doesburg JJ, De Vos L (1959) Pasteurized mixtures of fruit jus and milk, with long shelf life. V Int Fruit Jus Congr 20:32–37

Du B, Li J, Zhang H, Chen P, Huang L, Zhou J (2008) The stabilization mechanism of acidified milk drinks induced by carboxymethylcellulose. Lait 87(4–5):287–300. https://doi.org/10.1051/lait:2007021

Duan C, Meng X, Meng J, Khan MIH, Dai L, Khan A, An X, Zhang J, Huq T, Ni Y (2019) Chitosan as a preservative for fruits and vegetables: a review on chemistry and antimicrobial properties. J Bioresour Bioprod 4:11–21

FAO (2021) Riboflavins. Compendium of food additive specifications. In: 67th meeting 2006. FAO JECFA monographs 1. http://www.fao.org. Accessed 8 Jul 2021

Fardet A, Rock E (2017) In vitro and in vivo antioxidant potential of milks, yoghurts, fermented milks and cheeses: a narrative review of evidence. Nutr Res Rev 31:52–70

Gonelimali FD, Lin J, Miao W, Xuan J, Charles F, Chen M et al (2018) Antimicrobial properties and mechanism of action of some plant extracts against food pathogens and spoilage microorganisms. Front Microbiol 9:1639. https://doi.org/10.3389/fmicb.2018.01639

Haug M, Laubach C, Burke M, Harzer G (1987) Vitamin E in human milk from mothers of preterm and term infants. J Pediatr Gastroenterol Nutr 6(4):605–609

Hwang IG, Kim HY, Woo KS, Lee J, Jeong HS (2011) Biological activities of Maillard reaction products (MRPs) in a sugar-amino acid model system. Food Chem 126:221–227

Jeszka W, Jadwiga (2006) Food colorants, chemical and functional properties of food components. CRC Press, Boca Raton. https://doi.org/10.1201/9781420009613

Kaplow M (1970) Commercial development of intermediate moisture foods. Food Technol 24:53–57

Khan IT, Nadeem M, Imran M, Ullah R, Ajmal M, Jaspal MH (2019) Antioxidant properties of milk and dairy products: a comprehensive review of the current knowledge. Lipids Health Dis 18(1):41. https://doi.org/10.1186/s12944-019-0969-8

Krstonosic V, Jovicic-Bata J, Maravic N, Nikolic I, Dokic LJ (2021) Rheology, structure, and sensory perception of hydrocolloids. In: Food structure and functionality. Academic, New York, pp 23–47

Kumar N, Singh A, Sharma DK, Kishore K (2019) Toxicity of food additives. In: Food safety and human health. Elsevier, Amsterdam, pp 67–98

Leal-Calderon F, Schmitt V, Bibette J (2007) Emulsion science: basic principles, vol 3. Springer, New York, pp 173–199

Li N, Zhong Q (2020) Stable casein micelle dispersions at pH 4.5 enabled by propylene glycol alginate following a pH-cycle treatment. Carbohydr Polym 233:115834. https://doi.org/10.1016/j.carbpol.2020.115834

Lindmark-Månsson H, Åkesson B (2000) Antioxidative factors in milk. Br J Nutr 84:103–110

Litt MD, Duffy V, Oncken C (2016) Cigarette smoking and electronic cigarette vaping patterns as a function of e-cigarette flavourings. Tob Control 25:67–72

Mahmood K, Kamilah H, Shang PL, Sulaiman S, Ariffin F, Alias AK (2017) A review: interaction of starch/non-starch hydrocolloid blending and the recent food applications. Food Biosci 19: 110–120

Mahmud J, Khan RA (2018) Characterization of natural antimicrobials in food system. J Adv Microbial 8:11–894

Mani-Lopez E, Garcia HS, Lopez-Malo A (2012) Organic acids as antimicrobial to control *Salmonella* in meat and poultry products. Int Food Res J 45:713–721

Marta H, Cahyana Y, Djali M (2020) The effect of starch-hydrocolloid interaction on starch digestibility, pasting and physicochemical properties: a review. Earth Environ Sci 443(1): 012084. https://doi.org/10.1088/1755-1315/443/1/012084

McClements DJ (2015) Food emulsions principles, practices, and techniques, 3rd edn. CRC Press, Boca Raton, pp 314–407

Meenakshi S, Manicka DG, Tamil SM, Arumugam M, Balasubramanian T (2009) Total flavanoid and in vitro antioxidant activity of two seaweeds of Rameshwaram coast. Global J Pharmacol 3(2):59–62

Miranda PHS, dos Santos AC, de Freitas BCB, de Souza Martins GA, Boas EVDBV, Damiani C (2021) A scientific approach to extraction methods and stability of pigments from Amazonian fruits. Trends Food Sci Technol 113:335–345

Moneim A, Sulieman EH (2018) Gum arabic as thickener and stabilizing agents in dairy products. In: Gum arabic. Academic, New York. https://doi.org/10.1016/B978-0-12-812002-6.00013-0

Mostafavi FS, Zaeim D (2020) Agar-based edible films for food packaging applications - a review. Int J Biol Macromol 159:1165–1176. https://doi.org/10.1016/j.ijbiomac.2020.05.123. S0141813020332785

Motafavi FS, Zaeim D (2020) Agar-based edible films for food packaging applications—a review. Int J Biolog Macromol 159:1165–1176

Mussatto SI, Mancilha IM (2007) Non-digestible oligosaccharides: a review. Carbohydr Polym 68: 587–597

Najgebauer-Lejko D, Grega T, Tabaszewsa M (2014) Yoghurts with addition of selected vegetables: acidity, antioxidant properties and sensory quality. Acta Sci Pol Technol Aliment 13:35–42

Niki E (2014) Role of vitamin E as a lipid-soluble peroxyl radical scavenger: in vitro and in vivo evidence. Free Radic Biol Med 66:3–12

Nilsen-Nygaard J, Hattrem MN, Draget KI (2016) Propylene glycol alginate (PGA) gelled foams: a systematic study of surface activity and gelling properties as a function of degree of esterification. Food Hydrocolloids 57:80–91

Noisuwan A, Hemar Y, Wilkinson B, Bronlund JE (2009) Dynamic rheological and microstructural properties of normal and waxy rice starch gels containing milk protein ingredients. Starch-Starke 61:214–227

Park JW, Whiteside WS, Cho SY (2008) Mechanical and water vapor barrier properties of extruded and heat-pressed gelatin films. LWT Food Sci Technol 41(4):692–700. https://doi.org/10.1016/j.lwt.2007.04.015. S0023643807001752

Park SH, Lee CR et al (2020) Implications of agar and agarase in industrial applications of sustainable marine biomass. Appl Microbiol Biotechnol 104:2815–2832

Parker JK, Elmore S, Methven L (2014) Biosensors for food analysis. In: Parker JK, Elmore S, Methven L (eds) Flavour development, analysis and perception in food and beverages. Elsevier, pp 213–260

Peira G, Bollani L, Giachino C, Bonadonna A (2018) The management of unsold food in outdoor market areas: food operators' behaviour and attitudes. Sustainability 10(4):1180

Rai M, Pandit R, Gaikwad S, Kövics G (2016) Antimicrobial peptides as natural bio-preservative to enhance the shelf-life of food. J Food Sci Technol 53(9):3381–3394

Rehman A, Tong Q, Jafari SM, Assadpour E, Shehzad Q, Aadil RM, Iqbal MW, Rashed MM, Mushtaq BS, Ashraf W (2020) Carotenoid-loaded nanocarriers: a comprehensive review. Adv Colloid Interface Sci 275:102048

Rodriguez-Amaya DB (2015) Food carotenoids: chemistry, biology and technology. John Wiley & Sons

Rodriguez-Amaya DB (2016) Natural food pigments and colorants. Curr Opin Food Sci 7:20–26

Saha S (2020) To enhance the shelf-life of food. Int j Rec Trends Sci Technol 6(10):103–108

Saha D, Bhattacharya S (2010) Hydrocolloids as thickening and gelling agents in food: a critical review. J Food Sci Technol 47(6):587–597

Shan B, Cai YZ, Brooks JD, Corke H (2007) Antibacterial properties and major bioactive components of cinnamon stick (*Cinnamomum burmannii*): activity against foodborne pathogenic bacteria. J Agric Food Chem 55(14):5484–5490

Sharma M, Usmani Z, Gupta VK, Bhat R (2021) Valorization of fruits and vegetable wastes and by-products to produce natural pigments. Crit Rev Biotechnol 41:1–42

Siburian WZ, Rochima E, Andriani Y, Praseptiangga D (2020) Fish gelatin (definition, manufacture, analysis of quality characteristics, and application): a review. Int J Fish Aquat 8(4):90–95. E-ISSN: 2347-5129 P-ISSN: 2394-0506 (ICV-Poland)

Sigurdson GT, Tang P, Giusti MM (2017) Natural colorants: food colorants from natural sources. Annu Rev Food Sci Technol 8:261–280

Sinki G, Gordon R (2001) Flavoring agents. Food Sci Technol 13:47–51

Smith RL et al (2005a) GRAS flavoring substances 22. Food Technol 59:24–62

Smith RL, Cohen SM, Doull J, Feron VJ, Goodman JI, Marnett LJ, Portoghese PS, Waddell WJ, Wagner BM, Hall RL, Higley NA, Lucas-Gavin C, Adams TB (2005b) A procedure for the safety evaluation of natural flavor complexes used as ingredients in food: essential oils. Food Chem Toxicol 43(3):345–363. https://doi.org/10.1016/j.fct.2004.11.007. S0278691504003333

Song BJ, Jouni ZE, Ferruzzi MG (2013) Assessment of phytochemical content in human milk during different stages of lactation. Nutrition 29:195–120

Souza Mesquita LM, Martins M, Pisani LP, Ventura SPM, Rosso VV (2021) Insights on the use of alternative solvents and technologies to recover bio-based food pigments. Compr Rev Food Sci Food Saf 20(1):787–818. https://doi.org/10.1111/1541-4337.12685

Suetsuna K, Ukeda H, Ochi H (2000) Isolation and characterization of free radical scavenging activities peptides derived from casein. J Nutr Biochem 11:128–131

Tandjawa GA, Garnec CL, Boulenguer P, Gilles M, Langendorff V (2017) Rheological behavior of starch/carrageenan/milk proteins mixed systems: role of each biopolymer type and chemical characteristics. Food Hydrocolloids 73:300–312

Teigiserova DA, Hamelin L, Thomsen M (2020) Towards transparent valorization of food surplus, waste and loss: clarifying definitions, food waste hierarchy, and role in the circular economy. Sci Total Environ 706:136033

Tumbarski Y, Petkova N, Todorova M, Ivanov I, Deseva I, Mihaylova D, Ibrahim SA (2020) Effects of pectin-based edible coatings containing a bacteriocin of *Bacillus methylotrophicus* BM47 on the quality and storage life of fresh blackberries. Ital J Food Sci 32(2). https://doi.org/10.14674/IJFS-1663

Unsal YE, Soylak M, Tuzen M (2015) Ultrasound-assisted ionic liquid-based dispersive liquid–liquid microextraction for preconcentration of patent blue V and its determination in food samples by UV–visible spectrophotometry. Environ Monit Assess 187(4):1–8

Varela MS, Navarro AS, Yamul DK (2016) Effect of hydrocolloids on the properties of wheat/potato starch mixtures. Starch-Stärke 68:753–761

Vermeulen A, Nielsen DS, Marving CL, Daelman J (2015) Strategies to increase the stability of intermediate moisture foods towards *Zygosaccharomyces rouxii:* the effect of temperature, ethanol, pH and water activity, with or without the influence of organic acids. Food Microbiol 45:119–125

Vishal A, Aashima S, Kumari R, Vaishali S, Nidhi R, Bhatt AK (2019) In-vitro and in-silico characterization of xylose reductase from Emericella nidulans. Curr Chem Biol 13:159–170

Wang C, Wu J, Gao X (2015) Basic attributives, functions and characteristics of food additives. China Food Additiv 10:154–158

Wu L, Zhang Q, Shan L, Chen Z (2013a) Identifying critical factors influencing the use of additives by food: enterprises in China. Food Control 31:425–432

Wu J, Liu J, Dai Q, Zhang H (2013b) The stabilisation of acidified whole milk drinks by carboxymethylcellulose. Int Dairy J 28(1):40–42

Xing Y, Xu Q, Li X, Che Z, Yun J (2012) Antifungal activities of clove oil against *Rhizopus nigricans, Aspergillus flavus* and *Penicillium citrinum* in vitro and in wounded fruit test. J Food Saf 32(1):84–93

Yadav H, Maiti S (2020) Research progress in galactomannan-based nanomaterials: synthesis and application. Int J Biol Macromol 163:2113–2126. https://doi.org/10.1016/j.ijbiomac.2020.09.062. S0141813020344226

Yang Y, Fang Z, Chen X et al (2017) An overview of pickering emulsions: solid particle materials, classification, morphology, and applications. Front Pharmacol 8:287. https://doi.org/10.3389/fphar.2017.00287

Zeb A (2020) Concept, mechanism, and applications of phenolic antioxidants in foods. J Food Biochem 44(9):e13394. https://doi.org/10.1111/jfbc.13394

Zhang H, Chen J, Li J, Yan L, Li S, Ye X, Liu D, Ding T, Linhardt RJ, Orfila C et al (2018) Extraction and characterization of RG-I enriched pectic polysaccharides from mandarin citrus peel. Food Hydrocolloids 79:579–586

Chapter 9
Hypersensitivity Associated with Food Additives

Soniya Goyal, Mahiti Gupta, Pooja Sharma, and Vikas Beniwal

Abstract Food additives are substances added to packaged and processed foods to improve or maintain their freshness, texture, taste, or appearance. A wide range of additives is classified according to their functions, such as preservatives, coloring dyes, antioxidants, flavor enhancers, and sweeteners approved by FDA regulatory authorities, although some food additives have been associated with adverse reactions or hypersensitivities in many individuals. Adverse food reactions include immune-mediated allergic reactions and non-immune-mediated intolerance reactions. Both types of hypersensitivities are associated with symptoms related to skin, gastrointestinal, digestive, and respiratory systems such as urticaria, asthma, dermatitis, and anaphylaxis. The diagnosis of these adverse reactions is challenging and may be achieved by carefully analyzing clinical history followed by eradication and re-exposure with suspected food chemicals. For acute-onset allergic reactions, the determination of food-specific IgE antibodies is usually used for confirmation. Recent developments in diagnosing IgE-mediated adverse reactions related to food additives include single allergen assays, but the benchmark remains an oral food challenge. The best strategic approach for the management of food allergens is still its complete avoidance. It is mandatory that individuals with confirmed food allergies acquire optimal nutritional and dietetic assistance to manage their conditions. Thus, it is the responsibility of industries manufacturing food products to mention complete information regarding ingredients of food additives on the label of food products. This chapter mainly focuses on food additives involved in hypersensitivities.

Keywords Hypersensitivity · Food additives · Food allergy · Food intolerance · Food poisoning · Preservatives

S. Goyal · M. Gupta · P. Sharma
Department of Biotechnology, Maharishi Markandeshwar (Deemed to be University), Ambala, India

V. Beniwal (✉)
Department of Microbiology, Central University of Haryana, Mahendergarh, Haryana, India

A. K. Nadda, G. Goel (eds.), *Microbes for Natural Food Additives*, Microorganisms for Sustainability 38, https://doi.org/10.1007/978-981-19-5711-6_9

9.1 Introduction

Food additives are those substances integrated into edible food products to perform a range of specific functions, including preservation of food, thus increasing shelf life and making food's taste or texture better. According to European Legislation, a food additive is defined as any substance which is added to food intentionally for technological reasons in the production, processing, preparation, treatment, packaging, and storage of these food products, not normally consumed as the food itself and not normally used as a characteristic ingredient of food (Jen and Chen 2017). According to the FDA, food additives may be used for three main purposes: to improve or maintain safety and freshness, second to maintain the nutritional value, and third to maintain taste and texture. Food additives may be natural in origin as found to be an indigenous component of the food or may be artificial in origin as they are synthetically produced, such as azodicarbonamide, a flour improver which is used for holding bread dough together. Many natural additives such as salt, spices, sugar, and vinegar have been used from ancient times to preserve food products and make them more appetizing. Whether natural or synthetic, food additives are designated by their specific name or by a number, which in Europe is preceded by the letter E. The Romans used saltpeter (potassium nitrate) or turmeric as coloring agents to improve the appearance of certain products. Due to the increased demand of processed foods in the twentieth century, there came a need for new types and more use of food additives. Many advanced products, such as low-calorie and ready-to-eat snacks, would not be possible without the addition of food additives. Food additives are categorized into four categories: nutritional additives, preservatives, processing agents, and sensory agents. A list of different food additives given by the FDA can be diversified into different groups based on their specific purpose and function (Wilson and Bahna 2005) (Table 9.1).

- Anticaking: food additives used as anticaking agents prevent lumps in powders and crystalline substances.
- Chelating: food additives used as chelating agents bind with minerals present in food to prevent deterioration.
- Conditioning: compounds used as conditioning improve the baking quality of the flour.
- Bleaching: compounds used as bleaching agents enhance the color of food.
- Humectants: compounds used as humectants help in moisture control of packaged foods and keep them moist.
- Emulsifying: compounds used as emulsifying agents give good texture and homogeneity to food items by mixing immiscible items such as water and oil without any separation, for example ice creams and mayonnaise.
- pH control: food additives such as citric acid, lactic acid, and vinegar act as acid regulators, which control the acidic and basic nature of food and prevent them from spoilage.

Table 9.1 Different types of food additives used in different food products

Food additive class	Example	Present in	Role	References
Food preservative	BHA & BHT	Cereal, butter, dehydrated potatoes, chewing gum, beer	Provides stability and prevent discoloration in food	Botterweck et al. (2000)
Food fragrance	Octyl gallate	Bakery product	Fragrance and flavor	Gultekin and Dogue (2013)
Food color	Tartrazine	Soft drinks, chips, pudding, honey, pickles	Mixed with haldi to give color	Stevenson et al. (1992)
Food color	Sunset yellow	Pastry	Provides yellow color to food	Rajan et al. (2014)
Natural food color	Annatto	Ice creams, butter, cheeses, and bakery products	Provides red color to food	Ramsey et al. (2016)
Natural food color	Indigo carmine	Confectionary items, ice creams	Provides blue color to food	Magner and Gerebr (1994)
Flavoring agents	Peppermint oil	Chewing gums	Provides freshness and pungent taste	Bayat and Borici-Mazi (2014)
Flavor enhancer	Monosodium glutamate	Chinese food	Increases food palatability	Gultekin and Dogue (2013)
Emulsifiers	Lecithin	Chocolates and bakery	Stabilizes protein and fat emulsions, antioxidant flavor enhancer	Renaud et al. (1996)
Thickener, stabilizer, emulsifier	Guar gum	Ice creams	Binding agent and fixing agent	Mudgil et al. (2014)
Thickening and gelling agent	Locust bean gum	Jams and jellies	Stabilizes the consistency	Jedrzejczyk et al. (2020)
Preservatives	Sulfites	Wines, canned vegetables and fruits, dried fruits	Antibrowning agent and as antioxidant	Garcia-Fuentes et al. (2015)
Curing agents	Nitrate and nitrite	Meat and fish	Control growth of bacteria	Majou and Christieans (2018)
Laxative	Psyllium	Ice cream, biscuits, cereals	Slows gastric emptying and cholesterol-lowering source	Lambeau and McRorie (2017)
Artificial sweetener	Aspartame	Food and carbonated drinks	Noncalorie sweetener	Pang et al. (2021)

- Antioxidants and antimicrobial agents—these food additives prevent fat and oil rancidity in baked foods by hampering the effect of oxygen on food, thereby maintaining food palatability.
- Colorings: compounds used as coloring agents improve the appearance and texture of food, which attracts consumers.
- Flavorings and sweeteners: food additives also add flavor and sweetness.
- Stabilizers, gelling agents, and thickeners provide smoothness and strong texture to the food products.

Nowadays, different types of food additives have been commercially used worldwide, contributing to rapid growth in food processing and other related industries. The great success of food additives has fueled the development of new food processing technologies, resulting in unintended consequences that are a cause for concern. Despite all of the benefits of food additives and preservatives, there is still a risk of chemical adulteration of foods which may stimulate hormonal or chemical processes in the body resulting in undesirable physiological responses. Thus, not all food additives enter the market after being thoroughly tested to verify their safety. The residues produced by additives may also induce negative effects such as carcinogenic or toxic. Due to these reasons, most food additives are monitored and regulated by health authorities. All food manufacturers must meet the standards decided by the concerned authorities to assure the safety of final products consumed by the consumers. Food additives must go through a premarket safety evaluation to meet the standards with specific food additive regulations framed by specific government agencies such as the Food and Drug Administration (FDA) in the United States or European Food Safety Authority (EFSA) in Europe before using in foods (Rulis and Levitt 2009). Most food processing companies must get standard certification before introducing any new additive or preservative or using any originally certified preservative or additive differently.

Food additives may cause a wide range of adverse reactions. Some appear to be allergies, while others appear to be intolerances or sensitivities. Skin reactions, respiratory reactions, and gastrointestinal reactions (digestive reactions) are adverse reactions reported using food additives. There are approximately thousands of substances added to foods that are Generally Recognized As Safe (GRAS) by experts and are exempted from standard tolerance requirements (Neltner et al. 2011). Although some individuals are sensitive to certain food additives, particularly children, immunocompromised individuals, people with inherited metabolic disorders, and individuals with different xenobiotics metabolizing capacities, currently the American Academy of Pediatrics has raised concerns about the safety of GRAS in children (Trasande et al. 2018).

9.2 Hypersensitivity (Food Allergy)

A food allergy or hypersensitivity is an abnormal immune response to food or food additives. It is an adverse reaction toward food or food additives that can be mediated by two different mechanisms such as immunologic and nonimmunologic. A harmful physiological reaction caused by an immune system is termed as an allergy, whereas sometimes the symptoms appears to be similar to the allergic reaction but its mechanism is not truly allergic then it is known as intolerance. In intolerance conditions, mild symptoms will be observed, which are small enough to cause any toxic reactions. Hypersensitivity includes both allergic and intolerance conditions.

9.2.1 Classification and Mechanism of Hypersensitivity (Adverse Reactions)

Adverse food reactions (hypersensitivity) can be broadly classified into two reactions such as toxic and nontoxic reactions. Toxic compounds may have occurred naturally within a food, such as toxins present in rhubarb plant leaves or anthropogenic contaminants incorporated during food processing. Toxic reactions can occur in all exposed individuals with a sufficient dose, whereas nontoxic reactions depend on susceptibility to food in individuals. Nontoxic reactions may be mediated through two reactions such as immunological and nonimmunologic reactions (Fig. 9.1) (Onyimba et al. 2021). Immune-mediated reactions include food allergies and celiac diseases. Food allergies are defined as adverse immune responses to food or food

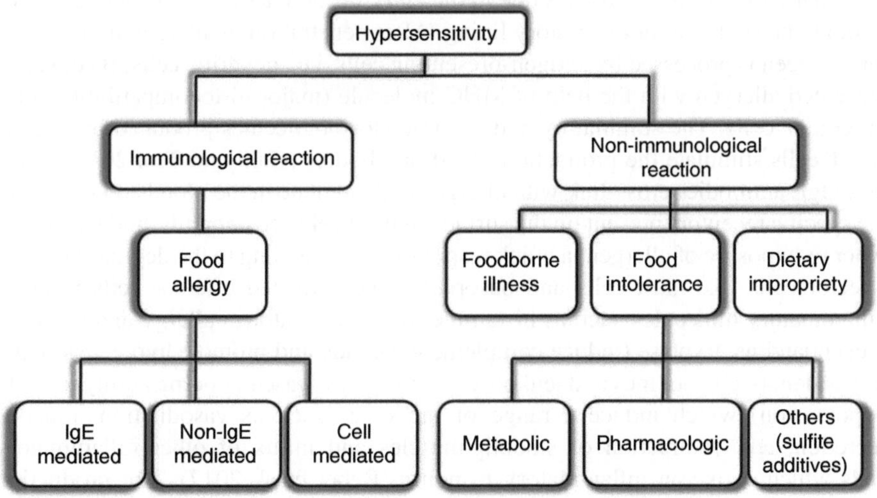

Fig. 9.1 Types of hypersensitivities developed through food and food additives

additives that cause typical clinical symptoms related to respiratory, gastrointestinal, cardiovascular, and neurological systems. Celiac disease is a disease associated with the digestive system and immune disorder which damages the small intestine and is triggered by gluten present in foods. This disease can cause long-term digestive issues and prevent an individual's body from getting all essential nutrients. Immune-mediated reactions can be further classified into three groups: IgE-mediated hypersensitivity, non-IgE-mediated hypersensitivity, and cell-mediated hypersensitivity (Montanez et al. 2017). On the other hand, nonimmunological reactions do not involve the immune system and are further classified into foodborne illness (food poisoning) and food intolerance. Food intolerance can be again divided into metabolic, pharmacologic, and undefined.

9.2.1.1 Immunologic Adverse Reactions to Food Additives

Depending on the type of immune response and involvement of immune cells to foreign antigens, immune-mediated reactions can be classified into two reactions such as IgE-mediated hypersensitivity and cell-mediated hypersensitivity. In IgE-mediated anaphylaxis, the mast cell is the main cell, while macrophages and neutrophils play an essential role in IgE-independent allergic reactions. Basophils are involved in both IgE-dependent and IgE-non-dependent allergic reactions (Moneret-Vautrin and Mertes 2010).

IgE-Mediated Hypersensitivity

IgE-mediated hypersensitivity, also known as type I hypersensitivity, occurs with the production of IgE antibodies due to the entry of food additives or food allergens through the skin, gut, or respiratory lining. After penetration of allergen in the body, this allergen is processed by antigen-presenting cells, i.e., dendritic cells, B cells, and presented allergen with the help of MHC molecule (major histocompatibility complex) to T cells. The stimulatory and costimulatory molecules produced by B cells and T cells stimulate the production of IgE antibodies (Galli and Tsai 2012). Then these IgE antibodies cross-link with allergen and circulate in the blood and bind with the specific receptors present on the surface of mast cells, eosinophils, and basophils. After re-exposure of allergen, a quick response arises, leading to the degranulation of effector cells like mast cells and basophils. The activated effector cells produce inflammatory molecules such as histamine (major mediator in allergy and asthma), prostaglandins, tryptase (induce complement cascade and promote hypotension and angioedema), chemokines and leukotrienes (increased vascular permeability leads to hypotension) which induce a range of processes such as vasodilation, mucous secretion, and contraction of smooth muscles and influx of other inflammatory cells which evoke an inflammatory response (Reber et al. 2017). The production of these potent chemical mediators by different types of cells is responsible for developing allergic symptoms.

A significantly less number of food additives induce IgE-mediated reactions but can be severe and life threatening. Natural additives are high molecular weight molecules sufficient to stimulate an IgE-mediated response. On the other hand, synthetic additives are low molecular weight compounds that mostly act like haptens. Such low molecular weight haptens after attachment to large carrier proteins can induce IgE-mediated immune response (Wilson and Bahna 2005). The generalized symptoms of IgE-mediated reactions are rapid in onset, leading to multisystematic manifestations, but generally, IgE-mediated reactions are considered acute reactions. Although late-phase reactions and recurrent exposures lead to the influx of inflammatory cells that are associated with chronic symptoms. The most prevalent symptoms are urticaria (hives) and angioedema, cutaneous manifestations (Ho et al. 2014).

Non-IgE-Mediated Hypersensitivity

There is another type of food allergy classified as a non-IgE-mediated food allergy caused by different immune system cells. This class of hypersensitivity includes two types of hypersensitivity: Type II (IgG/IgM-mediated) and Type III (IgG-mediated) hypersensitivity. This reaction is associated mainly with skin and digestive system symptoms such as heartburn, eczema, and digestion. In babies, this type of hypersensitivity is also associated with diarrhea and reflux (leakage of stomach acid into the throat). Type II hypersensitivity is antibody-mediated, resulting in host cell lysis by secreting toxic mediators. Two different mechanisms are involved in these cytotoxic reactions such as complement-mediated hypersensitivity (antibodies react with cell membrane components and activate the complement cascade, which leads to the lysis of cell) and antibody-dependent cell-mediated cytotoxicity (IgG and IgM Fc receptors are expressed on many cells such as macrophages, neutrophils, eosinophils, and natural killer cells and come in contact with antibody-coated target cells and lead to the lysis of cell by releasing cytoplasmic granules containing granzymes and perforin). The antibody-antigen complex mediates type III hypersensitivity. These circulating immune complexes may bind at various tissue sites and activate the complement cascade, resulting in tissue cell lysis. The accumulation of these complexes can lead to a hypersensitivity reaction that triggers various inflammatory processes. IgG antibody is involved in this type of reaction and does not bind with mast cells (Sole et al. 2021).

Cell-Mediated Hypersensitivity (Type IV Hypersensitivity Reaction)

Cell-mediated hypersensitivity is a delayed type of hypersensitivity triggered by antigen-specific T cells (T_H cells, primarily T_H1 subtype, and in few cases T_C cells). In this type, antibodies do not play any role. After exposure to the allergen, T_H1 cells get activated and release cytokines, which accumulate and activate macrophages resulting in local damage.

9.2.1.2 Nonimmunologic Adverse Reactions to Food Additives

Food Intolerance

Food intolerance is generally used for nonimmunological reactions, where the immune system does not play any role towards any food or food additives. These reactions are associated with direct stimulation of mast cells, which promote degranulation and release of mediators such as histamine, TNF-α, and prostaglandins, which trigger anaphylaxis or allergic reactions (Navines-Ferrer et al. 2018; Munoz-Cano et al. 2017). In this type, the spectrum of chemicals present in food and the quantity required to stimulate symptoms vary from individual to individual. Both naturally occurring food chemicals and food additives (natural or synthetic) cause an adverse effect on affected individuals. Most of the intolerance reactions are caused by salicylates and amines (the largest class of natural food substances), which affect the skin, gastrointestinal tract, central nervous system, and respiratory tract. Symptoms like irritable bowel, migraine, fatigue, asthma, urticaria, and some behavioral problems are also associated with intolerance to food additives. Sulfites, sodium benzoate, and food colors, among food additives, also show intolerance reactions in some patients. Food intolerance reactions may be metabolic and pharmacologic, and the mechanisms linked with intolerance reactions are still unknown (Sampson 2014).

Metabolic Intolerance

Metabolic intolerance related to food or food additives can be due to metabolic enzyme deficiencies, resulting in the inability to digest particular food or food additive. Lactose intolerance is one example of metabolic intolerance caused by lactase deficiency genetically and leads to indigestion of lactose. Lactose, mainly found in milk and milk-related products, is broken into the simplest sugars with the help of lactase which is vital for the nourishment of babies. Lactose intolerance causes a wide range of symptoms such as acute gastroenteritis, injury to the small intestine, bloating, cramps, and diarrhea due to lactic acid, carbon dioxide, and hydrogen gas production. Glucose-6-phosphate dehydrogenase deficiency is another example of metabolic intolerance that leads to the destruction of erythrocyte membranes (Paula Neto et al. 2017).

Pharmacological Intolerance

It was reported that the main class of substances present in food or food additives responsible for stimulating pharmacological food intolerance are salicylates and biogenic amines. The oral administration of salicylates (aspirin) for analgesic purposes causes side effects such as bronchial asthma and rhinitis in some individuals.

The high concentration of naturally occurring salicylates is present in vegetables, fruits, herbs, spices, and wines. Biogenic amines (low molecular weight nitrogenous compounds) are formed by decarboxylation of free amino acids or by transamination reactions during metabolism in plants, animals, and microorganisms. Biogenic amines are present in all food products such as fish products, dairy products, meat products, fermented vegetables, soy products, and beverages (wine and beer) which contain free amino acids and proteins. These amines have some toxic effects such as nausea, headaches, hypotension, hypertension, respiratory distress, along with other symptoms. Hayder et al. (2011) studied food additives such as ascorbic acid, MSG, tartrazine, sodium benzoate, and sodium metabisulfite using platelet activation test. These food additives inhibit platelet aggregation and are associated with the inhibition of the cyclooxygenase-thromboxane pathway. Aspirin interferes in the production of prostaglandins and leukotrienes and acts as a bronchodilator (Hannuksela and Haahtela 1987).

Unknown Mechanism Linked with Intolerance

In general, adverse reactions linked with food additives fall into this category in which not any specious biological mechanism has been proposed. The Australia New Zealand Food Standards Code strictly regulated the use of food additives based on technological requirements and ingredients present in them to ensure consumer safety. This code also designates the type and amount of additives used in what type of food. The most common food additives such as sulfite, sodium benzoate, and food coloring agents cause patients food intolerance reactions (Worm 2011).

Both IgE-mediated allergy and food additive intolerance reactions are mast cell-dependent reactions (Krystel-Whittemore et al. 2016). The activated mast cell releases histamines and other inflammatory mediators such as leukotrienes which lead to the onset of various clinical symptoms such as urticaria (skin allergy), redness of the skin, angioedema (if the allergen activates mast cells present in deeper tissue) as well as other organ-related symptoms like hypotension, dizziness, and dyspnea. In IgE-mediated response, cross-linking of membrane-bound IgE with allergen induces mast cell degranulation, while in the case of food intolerance reaction, additive or allergen directly induces mast cell activation. However, the exact mechanism is still not clear. Most adverse food reactions clinically seen in adults are associated with nonallergic hypersensitivity. For several years, many persons have reported adverse reactions to a range of food additives such as aspartame (an artificial sweetener), nitrates/nitrites (preservatives), monosodium glutamate (flavor enhancer), and sulfur based compounds (preservatives).

Food Poisoning

Food poisoning is a foodborne illness caused by eating contaminated food or toxins present in food. Food gets contaminated with infectious organisms at any step of processing, production, storing, or shipping. The symptoms related to food poisoning such as nausea, headache, vomiting, abdominal cramps, pain, fever, or diarrhea in individuals can start within hours of consuming contaminated products. Most symptoms associated with food poisoning are mild and disappear without any treatment, but individuals need to be hospitalized with food poisoning symptoms in some cases. Food poisoning especially occurs in the case of raw, ready-to-eat foods such as salads and other products because harmful organisms are not destroyed without cooking (Hernandez-Cortez et al. 2017).

9.2.2 Hypersensitivity Linked with Different Food Additives

Hypersensitivity reactions associated with different classes of food additives are mentioned below (Table 9.2).

9.2.2.1 Coloring Agents (Food Dyes)

Food colors are usually incorporated to food, beverages, cosmetics, and medications to make them more attractive and to enhance their color. It was found that natural food coloring agents are associated with many physiological dysfunctions in the body and produce some adverse reactions. The food products containing food colors consumed mostly by children are beverages (soft drinks) and sugar-based confectionery. The quantity of food colors daily consumed by individuals leads to the occurrence of adverse reactions. It is widely believed that food colors used in food industries to make existing food products brighter can trigger allergic (immune-mediated reactions) and food intolerance reactions (nonimmune reactions). Allergic reactions show a wide range of symptoms, from urticaria and asthma to anaphylaxis. Food colors can be classified as haptens (small low molecular weight molecules covalently attached with a carrier protein that can elicit an immune response). Both types of responses, i.e., IgE-mediated and non-IgE-mediated response, lead to histamine and leukotriene production, which results in the onset of typical allergy symptoms. The major food dyes used as coloring agents in food products responsible for food allergies are mentioned below.

9.2.2.2 Carmine

It is the most widely used biogenic dye in various products such as burgers, candies, soft drinks, some fruit yogurts, and cosmetics. The committees including Joint FAO/WHO expert committee on food additives (JECFA) and EU Scientific Committee for Food (SCF) have normalized the acceptable daily intake (ADI) for carmine as 5 mg/kg body weight (EFSA 2015). It is also found in red meat as the main food dye. Those individuals having beef allergies could be allergic to carmine

Table 9.2 A list of different food additives associated with different types of hypersensitivities

Food additive	Symptoms	Category	Mechanism	References
BHA and BHT	Red itchy skin, urticaria, cancer	Antioxidant	IgE mediated	Metcalfe et al. (2011)
Gallates	Pruritus, redness, and erythema	Food fragrance	Histamine	Gultekin and Dogue (2013)
Tartrazine	Anxiety, migraines, clinical depression, impaired senses, itching	Food color	IgE and non IgE mediated	Stevenson et al. (1992)
Sunset yellow	Asthma	Food color	Histamine	Rajan et al. (2014)
Annato	Urticaria	Natural food color	IgE-mediated sensitivity	Ramsey et al. (2016)
Indigo carmine	Urticaria	Natural food color	IgE-mediated sensitivity	Magner and Gerebr (1994)
Anethole	Sensitization in the oral cavity	Flavoring agents	Interleukins (IL-3, IL-6)	Bergau et al. (2021)
MSG	Headaches, nonrhythmic heartbeats, neurotoxicity	Flavor enhancer	B lymphocyte elicitor	Das et al. (2022)
Lecithin	Rhinitis, asthma, and whooping cough	Emulsifier generally in soy	Inflammation with release of histamines	Renaud et al. (1996)
Guar gum	Anaphylactic shock	Thickening agent	Basophil activation	Mudgil et al. (2014)
Locust bean gum	Vomiting, nausea, loose stools, diarrhea	Stabilizer	Food intolerance	Jedrzejczyk et al. (2020)
Benzoates	Urticaria and asthma	Preservative	IL-4 production by mononuclear cells	Yilmaz and Karabay (2018)
Mannitol	Anaphylactic shock	Sweetener	IgE-mediated mechanism (hapten binding to proteins)	Saha and Racine (2011)
Psyllium	Urticaria and anaphylaxis	Laxative	Cell-mediated IgA switched to IgE	Lambeau and McRorie (2017)

also. It has been associated with adverse reactions such as hives, angioedema, asthma, and anaphylaxis in adults (Chung et al. 2001; Tabar-Purroy et al. 2003; Kagi et al. 1994; Miyakawa et al. 2017). It is suggested that the IgE-mediated hypersensitivity mechanism is responsible for adverse reactions related to carmine containing products, resulting in rapid onset of symptoms such as cough, wheezing, hives, nausea, diarrhea, dermatitis, angioedema, and life-threatening anaphylaxis. Among these symptoms, immediate vomiting is the most common gastrointestinal symptom for IgE-mediated reactions (Onyimba et al. 2021). The different diagnostic tests such as skin prick test, basophil histamine release, and inhalational and oral challenges have shown that an IgE-mediated mechanism may be responsible for adverse reactions (Kagi et al. 1994; Beaudouin et al. 1995; Park 1981; Takeo et al. 2018).

9.2.2.3 Tartrazine

It is an approved artificial food coloring agent widely used in cosmetics, pharmaceuticals, and food products and gives a yellow color to products. The common symptoms of allergic reactions caused by tartrazine include urticaria (hives) and asthma. The Joint FAO/WHO expert committee on food additives (JECFA) and EU Scientific Committee for Food (SCF) have normalized the acceptable daily intake (ADI) for tartrazine as 7.5 mg/kg body weight (Rovina et al. 2017). ADI value indicates the amount of additives consumed daily without posing any significant risk to consumer health. Although these values standardized by regulatory authorities cannot completely eliminate the risk of adverse reactions associated with a specific substance, particularly for hypersensitive individuals or susceptible populations, they can significantly reduce the risk (Amchova et al. 2015), tartrazine adversely affects and alters biochemical markers in the liver and kidney at higher doses and even at low doses (Amin et al. 2010). Moneret-Vautrin (1983) experimentally proved the existence of IgE antibodies against this dye but were not observed by other investigators. Weliky et al. (1979) reported IgD antibodies specific to tartrazine although some investigators reported the presence of IgG antibodies against tartrazine (act as a hapten) conjugated with a protein carrier (Johnson et al. 1971).

9.2.2.4 Annatto

It gives yellow or orange color to food products. Annatto is mainly used as a natural food coloring agent in processed foods, beverages, and cheese. The adverse reactions related to this dye are mild and does not cause any severe symptoms. The IgE-mediate negative immune response has been reported due to annatto food color. This additive contains a large protein that can induce an immunological response that is associated with symptoms such as anaphylaxis and urticaria (Nish et al. 1991; Mikkelsen et al. 1978; Wilson and Bahna 2005). Theoretically, exposure to this additive can be through ingestion, inhalation, or skin, but in the case of children, ingestion is the main route of exposure. In an adult, IgE-mediated response to Annatto food color has been observed using basophil activation test (BAT) and IgE immunoblot tests (Stevens et al. 2014). The Joint FAO/WHO expert committee on food additives (JECFA) and EU Scientific Committee for Food (SCF) have normalized the acceptable daily intake (ADI) for annatto extracts such as Annatto bixin as 6 mg/kg body weight and Annatto norbixin as 0.3 mg/kg body weight (EFSA 2016).

9.2.2.5 Antioxidants and Antimicrobial Agents

Antioxidants such as BHT (butylated hydroxyl toluene) and BHA (butylated hydroxyl anisole) are used to prevent the spoilage of fats and oils. These antioxidants may cause adverse reactions such as hives and angioedema (swelling of deep layers of skin). Some investigators observe a delayed mechanism of hypersensitivity (cell-mediated hypersensitivity) in antioxidant and antimicrobial agents. The sensitization reactions may develop after skin contact with an allergen. Sodium nitrite is also used as an antioxidant and antimicrobial agent in different foodstuffs such as biscuits, meat products, and carbonated drinks. The Joint FAO/WHO expert committee on food additives (JECFA) and EU Scientific Committee for Food (SCF) have normalized the acceptable daily intake (ADI) for sodium nitrite as 0.2 mg/kg body weight. The intake of more than ADI value leads to the onset of symptoms such as hives, headache, or intestinal disorders. Sodium nitrite in some cases also enhances the effect of histamine present in many food items (Henderson and Raskin 1972).

9.2.2.6 Preservatives

The different types of preservatives such as nitrites (sodium nitrite), sulfites (sulfur dioxide, sodium sulfite), sorbates (sodium sorbates, potassium sorbates), and benzoate (sodium benzoate, benzoic acid) are used in a wide range of food products such as frozen mushroom, carbonated drinks, sauces, juices, low sugar products, cereal grains, and meat products for prevention of food spoilage (Dar et al. 2017). BHA and BHT are also the main preservatives used because of their antioxidant capacity. There are rare studies about adverse reactions to BHA and BHT in children. However, sulfites are associated with urticaria, anaphylaxis, and bronchoconstriction. Sulfites-related adverse reactions occur more often in asthmatic patients. In adults, adverse reactions associated with preservatives include contact dermatitis, abdominal cramps, diarrhea, and bronchoconstriction (Laura et al. 2019). Sodium metabisulfite is associated with urticaria and angioedema. In some cases, IgE-mediated immune response mediates adverse reactions in sulfites along with several other mechanisms. One possible mechanism is that sulfites in food and beverages are converted into sulfur dioxide by gastric acidity, which evokes smooth bronchial muscle contraction. Another mechanism is that sulfur dioxide indirectly stimulates cholinergic reflex, which stimulates bronchoconstriction (Wilson and Bahna 2005).

9.2.2.7 Emulsifiers and Stabilizers

Lecithin is used as an emulsifier in many food products such as soybeans, eggs, rice, sunflower seeds, and rapeseed. It is a mixture of phosphatides. Lecithin contains phospholipids, soy protein, and soy allergen residues (Nicolson and Settineri 2021).

Adverse reactions associated with soy lecithin have been described in a few cases (Mortensen et al. 2017). A disease commonly known as Baker's asthma is seen among workers of the baking industry. Lavaud et al. (1994) revealed the occurrence of rhinitis, asthma, and cough with sputum in individuals working in the baking industry. The positive results of the skin prick test showed the presence of soy lecithin. Individuals associated with these symptoms were given a soy lecithin-free diet which then improved their symptoms (Renaud et al. 1996). The different types of gums such as guar gum, locust bean, alginate, and acacia are also used in foods as stabilizers. However, allergic reactions associated with ingestion of gums are infrequent. Papanikolaou et al. (2007) found a case of anaphylaxis with the ingestion of guar gum in several foods and beverages. Adverse reactions associated with guar gum are mediated through basophil activation. Bridts et al. (2002) reported symptoms such as severe contact urticaria through guar gum used for a local anesthetic in a dental procedure.

9.2.2.8 Flavorings and Taste Enhancers

Monosodium glutamate (MSG) is a flavor enhancer added to various foods, canned vegetables, soups, and meats. According to the US Food and Drug Administration (FDA), MSG is classified as GRAS. But its use is still controversial; FDA requires MSG to be listed on the label. The FDA has reported many adverse reactions related to the use of MSG known as MSG symptom complex in sensitive individuals who consume 3 g or more. This complex includes headache, sweating, face tightness, chest pain, nausea, and weakness but no clear proof of linkage was found. The symptoms are often mild because MSG foods do not contain more than 0.5 g. The only way of prevention is to avoid foods containing MSG (Gultekin and Dogue 2013). MSG adverse reactions are mainly associated with food intolerance reactions or IgE-independent reactions (Wilson and Bahna 2005). Spices are the major group of food additives used as flavoring agents that have been shown to evoke various adverse reactions such as immunological contact urticaria. Some spices also produced nonimmunological contact urticaria. Aspartame is used as a calorie-free sweetener in many foods and beverages. The adverse reactions associated with sweetener may include eye problems, decreased vision, hearing problems, headache, confusion, dizziness, psychological, chest pain, gastrointestinal issues, neurological damage, and skin allergies. High fructose corn syrup (HFCS) is used as an artificial sweetener in almost all processed foods, bread, flavored yogurt, canned vegetables, and cereals. HFCS may lead to adverse symptoms such as the development of diabetes, tissue damage, and increased LDL cholesterol levels.

9.2.3 *Prevalence of Hypersensitivity Reactions to Food Additives*

Few scientists have investigated the prevalence of adverse reactions to different food additives. It is reported in some studies that the prevalence of adverse reactions to food additives in adults is less than 1%, while in children, it seems to be higher (1–2%) (Fuglsang et al. 1993, 1994; Feketea and Tsabouri 2017). Many food additives are widely used in our food industries, and their adverse reactions to additives seem based on their low index of suspicion (Wilson and Bahna 2005). It was reported that the frequency of food additive intolerance is 0.03–0.15%, based on data from patient groups. In the British population, only food additive intolerance prevalence is estimated at nearly 0.026%. Similarly, in the Danish population, the prevalence of adverse reactions due to food additives is 1–2% in children aged 5–16 (Madsen 1994). Food additive intolerance is primarily found in atopic children with cutaneous symptoms where the additive is aggravating an existing disease. Food additive intolerance is also found in adults with atopic symptoms from the respiratory tract and skin. In adults and children, subjective symptoms are reproducible, with headache and behavior or mood swings with less prevalence (0.026%). In many cases, epidemiological studies related to food additive intolerance have been carried out on selected patients suffering from asthma, rhinitis, or urticaria (Cianferoni 2016; Simon et al. 2015; Anıl and Harmancı 2020). Juhlin calculated the prevalence of food additive intolerance in the Swedish population suffering from urticaria, angioedema, asthma, and hay fever (the most commonly identified reactions) and found a prevalence of 0.4% for aspirin intolerance, 0.6% for tartrazine reactions, and 0.5% for benzoate intolerance (Young et al. 1987).

9.2.4 *Diagnostic and Management Strategies of Hypersensitivities to Food Additives*

9.2.4.1 Biomarkers for Food Allergens

Allergies due to food are prevalent throughout the world today. It is mainly due to synthetic food additives used daily to enhance the taste and life span of the food. Various biomarkers can test the sensitivity of food allergies. Generally, allergies might be skin and gut related or stomach related (Fig. 9.2).

9.2.4.2 Immunomarkers

These markers elicit the immune response leading to activation of either B cells or T cells. B cells further produce various immunoglobulin, generally IgE. IgA and IgG

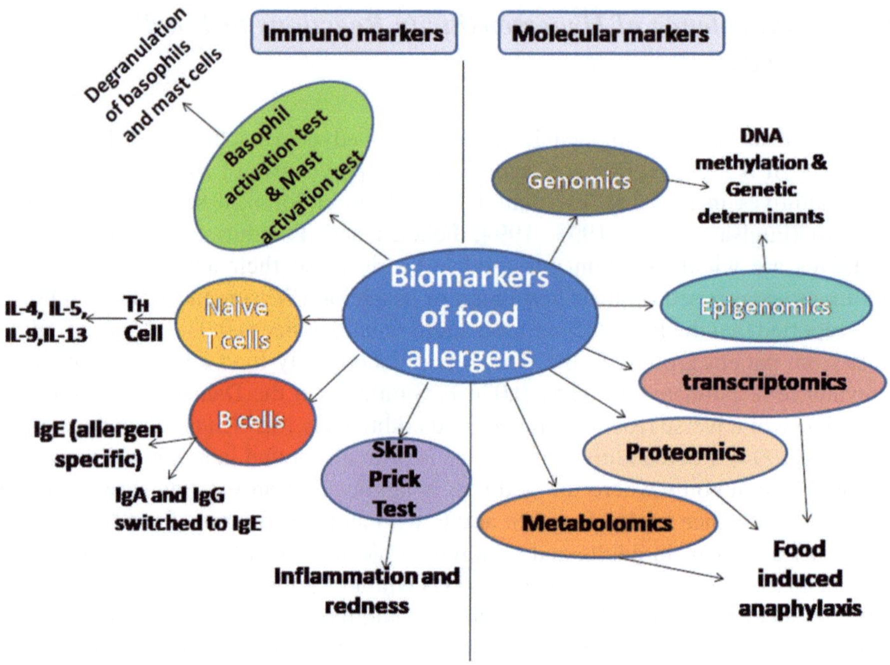

Fig. 9.2 Biomarkers as a diagnostic approach for the sensitivity of food allergens

are also produced, but they ultimately lead to the switching of IgE antibodies (Foong and Santos 2021).

9.2.4.3 Skin Prick Test (SPT)

In this test, skin is pricked and the allergen is injected inside the body at sublethal concentrations. If the reaction is positive, histamine release will lead to inflammation and redness.

9.2.4.4 Allergen-Specific T Cells/Specific Immunoglobulin Elicitation

Part or whole allergen is used to elicit specific immunoglobulin responses. Firstly the allergen interacts with antigen-presenting cells that elicit T_H cells which secrete cytokines. B cells start secreting IgE antibodies specific to the allergen in serum.

9.2.4.5 Basophil Activation Test (BAT)

BAT exploits the capacity of IgE immunoglobulins to elicit basophil cells that display certain marker domains CD63 and CD203 and have already been stimulated by a specific allergen.

9.2.4.6 Mast Cell Activation Test

It is quite similar to BAT, but the allergen epitope sensitizes LAD2 mast cells here. Instead of serum, human plasma is used for testing.

9.2.4.7 IgG/IgA

Food-specific IgG and IgA are not very good diagnoses for allergens as they do not show cross-reactivity.

9.2.4.8 Molecular Markers

Specific DNA methyl patterns and other profile studies can study changes induced by allergen at the DNA level and expression level (Patil et al. 2020).

9.2.4.9 Transcriptomics

It is an in vivo approach in which allergen is injected inside the body and peripheral blood samples are obtained before, during, and after interaction with the allergen. Potential changes are identified in DNA and expression during an allergic reaction.

9.2.4.10 Epigenomics

Allergens have different reactions in different individuals controlled by genes and the environment. Experts have well studied changes in the DNA methylation pattern of CD4 lymphocytes and CpG islands, leading to the profiling of changes produced by allergens.

9.2.4.11 Metabolomics

In this approach, the end products of metabolism of cellular processes are studied. The metabolites are studied using different analytical techniques like LCMS and

GCMS. The changes in metabolites due to reactions linked with allergen lead to the diagnosis of a particular change in metabolism.

The optimum management strategy for prevention of adversative reactions related to food additives is to strictly eliminate suspicious eatable additives from the diet. If patients have symptoms related to skin and the gastrointestinal system, they should take antihistamines in their diet. The diagnostic procedures include eliminating diet followed by double-blind-placebo-controlled-food challenge (DBPCFC) experiments. DBPCFC is one of the main diagnostic tests used in food allergies as it minimizes diagnostic bias (Cerecedo et al. 2014). In case the DBPCFC test is positive, further diagnosis can be verified and endorsements related to diet can be completed. Some of the preliminary tests such as skin prick test and specific IgE test could support the above diagnosis. Similarly, peak expiratory flows at and off work is widely acceptable by most clinicians to monitor asthma with a better diagnostic value. In IgE-mediated allergic reactions, total calculation of IgE level and additive specific IgE in blood sample is the highly effective diagnosis and management approach. Avoiding specific allergens responsible for inducing allergy gives quick relief to the allergy or asthma. The main issue during diagnosis related to food intolerance is that the patients might not be clear about their history of symptoms. In these situations, a symptom diary might be supportive. Generally in vivo tests (skin prick test) and in vitro test (detection of specific IgE antibodies) are not more effective for proper diagnosis (Reese et al. 2009). It is also found that insufficient knowledge and proper guidance of toxicology might result in adverse effects which may lead to death (Gram et al. 2002; Badora et al. 2019). In addition to this, food companies are mainly focusing on organoleptic properties of their products and are working only to enhance food with its flavors, colors, and sweeteners, without any focus on scientific proof of their effects on human health. In 1907. The United States evaluated 90 synthetic colors used in the food dyeing industry but only seven were found to be acceptable for further use. There should be requirements of comprehensive studies and strict regulations on food additives but regulations for food additives were made after a century (Badora et al. 2019; Wozniak et al. 2021). However later on food safety and security is ensured by the World Health Organization (WHO) and the Food and Agriculture Organization (FAO). In 1962, these organizations established a special agenda for food safety rules entitled "Codex Alimentarius Commission." The Commission has prepared for updating the Codex Alimentarius. This Act is not a legal Act but can offer a reference for minimal standards on raw resources and food products, contamination levels, hygienic processing, research methods, and food additives all over the world (Badora et al. 2019). Similarly, a European Union body such as European Food Safety Authority (EFSA) is responsible for improving human health, protection, and food safety/ security risk mitigation. This scientific agency was established in early 2002. European Union legislation has approved approximately 330 food additives for food industrial use (Badora et al. 2019). Similarly in Poland, the Food and Nutrition Safety Act (FNSA) was established in 2006 governing food safety rules. The FNSA is responsible for supplying food and nutrition products, labeling products, maintaining hygienic conditions during the course of production process, product replacement rules, and cercerns regarding the use of food additives.

The European Food Safety Authority (EFSA) and the WHO-FAO-JECFA are the supreme authorities who have the initial power to set the levels of food additives at international level. They analyze empirically and theoretically and highlight the potential adverse effects of food additives to protect consumers of a given food product. There is an Open Food Facts database from where one can retrieve the composition of any food product using the weblink http://world.openfoodfacts.org/. This Open Food Facts Database (OFFD) is an open, collaborative, and informative database for standard food products marketed worldwide, licensed under the Open Database License (ODBL). This is a French initiative project established in 2012 that contains information on thousands of food products. It is a good source of getting information regarding French food market and availability of different food products for other countries worldwide. A specific Global Trade Item Number (GTIN) is inserted within the barcode of each product (Chazelas et al. 2020). Furthermore, every food additive has its own standard code for identity, harmony, and good relation for future. The given code is specific and consistent with the International Numbering System (INS) standard.

9.3 Conclusion

In literature, food allergies and food intolerance reactions are widely used interchangeably and indiscriminately. Generally, food intolerance reactions are distinguished from food allergies based on their molecular mechanism and severity of symptoms. Their metabolic and physiological mechanisms are well documented in earlier studies. But no precise mechanisms are proposed for intolerance reactions associated with food additives. The medical experts suggest that food additives make patients' illnesses more likely to manifest as symptoms.

In summary, food additives are increasing continuously throughout the world day by day, and there are various reports on a wide range of severe symptoms associated with food additives. Although there is a lack of knowledge regarding causative agents, manifestations and mechanisms, regulation, and stringent examination of adverse reactions related to food additives, in some cases, it is suggested that asthma is associated with the consumption of sulfites and benzoates. Anaphylaxis is associated with food colors (annatto, carmine), sulfites, sweeteners, nitrites, and benzoates. Similarly urticaria may be due to consumption of antioxidants, aspartame, and MSG. Hence, diagnosing hypersensitivity reactions against food additives is a tedious and challenging task than diagnosing food protein allergy.

References

Amchova P, Kotolova H, Ruda-Kucerova J (2015) Health safety issues of synthetic food colorants. Regul Toxicol Pharmacol 73(3):914–922

Amin KA, Hameid HA II, Abd Elsttar AH (2010) Effect of food azo dyes tartrazine and carmoisine on biochemical parameters related to renal, hepatic function and oxidative stress biomarkers in young male rats. Food Chem Toxicol 48(10):2994–2999

Anıl H, Harmancı K (2020) Evaluation of contact sensitivity to food additives in children with atopic dermatitis. Adv Dermatol Allergol 37(3):390–395

Badora A, Bawolska K, Kozłowska-Strawska J, Domańska J (2019) Food additives in food products: a case study. In: Nutrition in health and disease—our challenges now and forthcoming time. IntechOpen, London

Bayat R, Borici-Mazi R (2014) A case of anaphylaxis to peppermint. Allergy Asthma Clin Immunol 10(1):6. https://doi.org/10.1186/1710-1492-10-6

Beaudouin E, Kanny G, Lambert H, Fremont S, Moneret-Vautrin DA (1995) Food anaphylaxis following ingestion of carmine. Ann Allergy Asthma Immunol 74(5):427–430

Bergau N, Herfurth UM, Sachse B, Abraham K, Monien BH (2021) Bioactivation of estragole and anethole leads to common adducts in DNA and hemoglobin. Food Chem Toxicol 153:112253

Botterweck AA, Verhagen H, Goldbohm R, Kleinjans J, van den Brandt P (2000) Intake of butylated hydroxyanisole and butylated hydroxytoluene and stomach cancer risk: results from analyses in the Netherlands Cohort Study. Food Chem Toxicol 38:599–605

Bridts CH, Ebo DG, De Clerck LS, Stevens WJ (2002) Anaphylaxis due to the ingestion of guar gum. J Allergy Clin Immunol 109(1):S221

Cerecedo I, Zamora J, Fox M, Voordouw J, Plana N, Rokicka E et al (2014) The impact of double-blind placebo-controlled food challenge (DBPCFC) on the socioeconomic cost of food allergy in Europe. J Investig Allergol Clin Immunol 24(6):418–424

Chazelas E, Deschasaux M, Srour B, Kesse-Guyot E, Julia C, Alles B et al (2020) Food additives: distribution and co-occurrence in 126,000 food products of the French market. Sci Rep 10(1): 1–15

Chung K, Baker JR Jr, Baldwin JL, Chou A (2001) Identification of carmine allergens among three carmine allergy patients. Allergy 56(1):73–77

Cianferoni A (2016) Wheat allergy: diagnosis and management. J Asthma Allergy 9:13

Dar HY, Chaturvedi S, Srivastava K, Azam Z, Anupam R, Mondal RK et al (2017) Immunomodulatory effects of food additives. Int J Immunother Cancer Res 3(1):019–031

Das D, Banerjee A, Bhattacharjee A, Mukherjee S, Maji BK (2022) Dietary food additive monosodium glutamate with or without high-lipid diet induces spleen anomaly: a mechanistic approach on rat model. Open Life Sci 17(1):22–31. https://doi.org/10.1515/biol-2022-0004

EFSA Panel on Food Additives and Nutrient Sources added to Food (ANS) (2015) Scientific opinion on the re-evaluation of cochineal, carminic acid, carmines (E 120) as a food additive. EFSA J 13(11):4288

EFSA Panel on Food Additives and Nutrient Sources added to Food (ANS) (2016) The safety of annatto extracts (E 160b) as a food additive. EFSA J 14(8):e04544

Feketea G, Tsabouri S (2017) Common food colorants and allergic reactions in children: myth or reality? Food Chem 230:578–588

Foong RX, Santos AF (2021) Biomarkers of diagnosis and resolution of food allergy. Pediatr Allergy Immunol 32(2):223–233

Fuglsang G, Madsen C, Saval P, Osterballe O (1993) Prevalence of intolerance to food additives among Danish school children. Pediatr Allergy Immunol 4(3):123–129

Fuglsang G, Madsen C, Halken S, Jorgensen M, Ostergaard PA, Osterballe O (1994) Adverse reactions to food additives in children with atopic symptoms. Allergy 49(1):31–37

Galli SJ, Tsai M (2012) IgE and mast cells in allergic disease. Nat Med 18(5):693–704

Garcia-Fuentes AR, Wirtz S, Vos E, Verhagen H (2015) Short review of sulphites as food additives. Eur J Nutr Food Saf 5(2):113–120

Gram L, Ravn L, Rasch M, Bruhn JB, Christensen AB, Givskov M (2002) Food spoilage—interactions between food spoilage bacteria. Int J Food Microbiol 78(1–2):79–97
Gultekin F, Dogue DK (2013) Allergic and immunologic reactions to food additives. Clin Rev Allergy Immunol 45:6–24
Hannuksela M, Haahtela T (1987) Hypersensitivity reactions to food additives. Allergy 42(8): 561–575
Hayder H, Mueller U, Bartholomaeus A (2011) Review of intolerance reactions to food and food additives. Int Food Risk Anal J 1(2):23–32
Henderson W, Raskin N (1972) "Hot-dog" headache: individual susceptibility to nitrite. Lancet 300(7788):1162–1163
Hernandez-Cortez C, Palma-Martínez I, Gonzalez-Avila LU, Guerrero-Mandujano A, Solís RC, Castro-Escarpulli G (2017) Food poisoning caused by bacteria (food toxins). In: Poisoning: from specific toxic agents to novel rapid and simplified techniques for analysis. IntechOpen, London, p 33
Ho MHK, Wong WHS, Chang C (2014) Clinical spectrum of food allergies: a comprehensive review. Clin Rev Allergy Immunol 46(3):225–240. J Dermatol 2017;44:96–7
Jedrzejczyk M, Bartnik K, Funkowicz M, Toporowska KE (2020) Locust bean gum induced FPIES in infant. J Investig Allergol Clin Immunol 30:2–3
Jen JJS, Chen J (eds) (2017) Food safety in China: science, technology, management and regulation. Wiley, New York
Johnson HM, Peeler JT, Smith BG (1971) Tartrazine: quantitative passive hemagglutination studies on a food-borne allergen of small molecular weight. Immunochemistry 8(3):281–287
Kagi M, Wüthrich B, Johansson SGO (1994) Campari-Orange anaphylaxis due to carmine allergy. Lancet 344(8914):60–61
Krystel-Whittemore M, Dileepan KN, Wood JG (2016) Mast cell: a multi-functional master cell. Front Immunol 6:620. https://doi.org/10.3389/fimmu.2015.00620
Lambeau KV, McRorie JW Jr (2017) Fiber supplements and clinically proven health benefits: how to recognize and recommend an effective fiber therapy. J Am Assoc Nurse Pract 29(4):216–223
Laura A, Arianna G, Francesca C, Carlo C, Carla M, Giampaolo R (2019) Hypersensitivity reactions to food and drug additives: problem or myth? Acta Biomed 90(Suppl 3):80
Lavaud F, Perdu D, Prcvost A, Vallcrand H, Cossart C, Passemard F (1994) Baker's asthma related to soybean lecithin exposure. Allergy 49:159–162
Madsen C (1994) Prevalence of food additive intolerance. Hum Exp Toxicol 13(6):393–399
Magner J, Gerebr P (1994) Urticaria due to blue dye in synthroid tablets. Thyroid 4(3):341
Majou D, Christieans S (2018) Mechanisms of the bactericidal effects of nitrate and nitrite in cured meats. Meat Sci 145:273–284
Metcalfe DD, Sampson HA, Simon RA, (Eds.). (2011) Food allergy: adverse reactions to foods and food additives. Wiley, New York
Mikkelsen H, Larsen JC, Tarding F (1978) Hypersensitivity reactions to food colours with special reference to the natural colour annatto extract (butter colour). In: Toxicological aspects of food safety. Springer, Berlin, pp 141–143
Miyakawa M, Inomata N, Sagawa N, Nomura Y, Yamaguchi Y, Aihara M (2017) Anaphylaxis due to carmine-containing foods induced by epicutaneous sensitization to red eye-liner. J Dermatol 44(1):96–97
Moneret-Vautrin DA (1983) False food allergies: non-specific reactions to foodstuffs. Clinical reactions to food, pp 135–153
Moneret-Vautrin A, Mertes PM (2010) Anaphylaxis to general anesthetics. Anaphylaxis 95:180–189
Montanez MI, Mayorga C, Bogas G, Barrionuevo E, Fernandez-Santamaria R, Martin-Serrano A et al (2017) Epidemiology, mechanisms, and diagnosis of drug-induced anaphylaxis. Front Immunol 8:614
Mortensen A, Aguilar F, Crebelli R, Di Domenico A, Frutos MJ, Galtier P, Gott D, Gundert-Remy U, Lambré C, Leblanc JC, Lindtner O, Moldeus P, Mosesso P, Oskarsson A,

Parent-Massin D, Stankovic I, Waalkens-Berendsen I, Woutersen RA, Wright M, Younes M, Brimer L, Altieri A, Christodoulidou A, Lodi F, Dusemund B (2017) Re-evaluation of lecithins (E 322) as a food additive. EFSA J 15(4):e04742. https://doi.org/10.2903/j.efsa.2017.4742

Mudgil D, Barak S, Khatkar BS (2014) Guar gum: processing, properties and food applications—a review. J Food Sci Technol 51(3):409–418. https://doi.org/10.1007/s13197-011-0522-x

Munoz-Cano R, Pascal M, Araujo G, Goikoetxea MJ, Valero AL, Picado C, Bartra J (2017) Mechanisms, cofactors, and augmenting factors involved in anaphylaxis. Front Immunol 8:1193

Navines-Ferrer A, Serrano-Candelas E, Lafuente A, Munoz-Cano R, Martín M, Gastaminza G (2018) MRGPRX2-mediated mast cell response to drugs used in perioperative procedures and anaesthesia. Sci Rep 8(1):1–11

Neltner TG, Kulkarni NR, Alger HM, Maffini MV, Bongard ED, Fortin ND, Olson ED (2011) Navigating the US food additive regulatory program. Compr Rev Food Sci Food Saf 10(6): 342–368

Nicolson G, Settineri R (2021) No evidence of allergenic reactions to soy lecithin phospholipids used in membrane lipid replacement studies. Bioactive Compd Health Dis 4(1):9–13

Nish WA, Whisman BA, Goetz DW, Ramirez DA (1991) Anaphylaxis to annatto dye: a case report. Ann Allergy 66(2):129–131

Onyimba F, Crowe SE, Johnson S, Leung J (2021) Food allergies and intolerances: a clinical approach to the diagnosis and management of adverse reactions to food. Clin Gastroenterol Hepatol 19(11):2230–2240

Pang MD, Goossens GH, Blaak EE (2021) The impact of artificial sweeteners on body weight control and glucose homeostasis. Front Nutr 7:598340

Papanikolaou I, Stenger R, Bessot JC, De Blay F, Pauli G (2007) Anaphylactic shock to guar gum (food additive E412) contained in a meal substitute. Allergy 62(7):822–822

Park GR (1981) Anaphylactic shock resulting from casualty simulation. A case report. J R Army Med Corps 127(2):85–86

Patil SU, Bunyavanich S, Berin MC (2020) Emerging food allergy biomarkers. J Allergy Clin Immunol Pract 8(8):2516–2524. https://doi.org/10.1016/j.jaip.2020.04.054

Paula Neto HA, Ausina P, Gomez LS, Leandro JG, Zancan P, Sola-Penna M (2017) Effects of food additives on immune cells as contributors to body weight gain and immune-mediated metabolic dysregulation. Front Immunol 8:1478

Rajan JP, Simon RA, Bosso JV (2014) Prevalence of sensitivity to food and drug additives in patients with chronic idiopathic urticaria. J Allergy Clin Immunol Pract 2(2):168–171

Ramsey NB, Tuano KTS, Davis CM, Dillard K, Hanson C (2016) Annatto seed hypersensitivity in a pediatric patient. Ann Allergy Asthma Immunol 117(3):331–333

Reber LL, Hernandez JD, Galli SJ (2017) The pathophysiology of anaphylaxis. J Allergy Clin Immunol 140(2):335–348

Reese I, Zuberbier T, Bunselmeyer B, Erdmann S, Henzgen M, Fuchs T et al (2009) Diagnostic approach for suspected pseudoallergic reaction to food ingredients. J Dtsch Dermatol Ges 7(1): 70–77

Renaud C, Cardiet C, Dupont C (1996) Allergy to soy lecithin in a child. J Pediatr Gastroenterol Nutr 22(3):328–329

Rovina K, Siddiquee S, Shaarani SM (2017) A review of extraction and analytical methods for the determination of tartrazine (E 102) in foodstuffs. Crit Rev Anal Chem 47(4):309–324

Rulis AM, Levitt JA (2009) FDA's food ingredient approval process: safety assurance based on scientific assessment. Regul Toxicol Pharmacol 53(1):20–31

Saha BC, Racine FM (2011) Biotechnological production of mannitol and its applications. Appl Microbiol Biotechnol 89(4):879–891

Sampson HA (2014) Food allergy: a winding road to the present. Pediatr Allergy Immunol 25(1): 25–26

Simon RA, Sicherer SH, Feldweg AM (2015) Allergic and asthmatic reactions to food additives

Sole D, Spindola MAC, Aun MV, Araújo Azi LMTD, Bernd LAG, Garcia DB et al (2021) Update on perioperative hypersensitivity reactions: joint document from the Brazilian Society of

Anesthesiology (SBA) and Brazilian Association of Allergy and Immunology (ASBAI)-part II: etiology and diagnosis. Rev Bras Anestesiol 70:642–661
Stevens LJ, Burgess JR, Stochelski MA, Kuczek T (2014) Amounts of artificial food colors in commonly consumed beverages and potential behavioral implications for consumption in children. Clin Pediatr 53(2):133–140
Stevenson DD, Simon RA, Lumry WR, Mathison DA (1992) Pulmonary reactions to tartrazine. Pediatr Allergy Immunol 3(4):222–227
Tabar-Purroy AI, Alvarez-Puebla MJ, Acero-Sainz S, García-Figueroa BE, Echechipía-Madoz S, Olaguibel-Rivera JM, Quirce-Gancedo S (2003) Carmine (E-120)–induced occupational asthma revisited. J Allergy Clin Immunol 111(2):415–419
Takeo N, Nakamura M, Nakayama S, Okamoto O, Sugimoto N, Sugiura S et al (2018) Cochineal dye-induced immediate allergy: review of Japanese cases and proposed new diagnostic chart. Allergol Int 67(4):496–505
Trasande L, Shaffer RM, Sathyanarayana S, Lowry JA, Ahdoot S, Baum CR et al (2018) Food additives and child health. Pediatrics 142(2):e20181410
Weliky N, Heiner DC, Tamura H, Anderson S, Stenius-aarniala B, German DF et al (1979) Correlation of tartrazine hypersensitivity with specific serum IgD levels. Immunol Commun 8(1):65–71
Wilson BG, Bahna SL (2005) Adverse reactions to food additives. Ann Allergy Asthma Immunol 95(6):499–507
Worm M (2011) Q&A: food additive intolerance. BMC Med 9(1):1–2
Wozniak E, Tyczewska A, Twardowski T (2021) Bioeconomy development factors in the European Union and Poland. New Biotechnol 60:2–8
Yilmaz B, Karabay AZ (2018) Food additive sodium benzoate (NaB) activates NFκB and induces apoptosis in HCT116 cells. Molecules (Basel, Switzerland) 23(4):723. https://doi.org/10.3390/molecules23040723
Young E, Patel S, Stoneham MD, Rona R, Wilkinson JD (1987) The prevalence of reaction to food additives in a survey population. J R Coll Physicians Lond 21(4):241–247

MIX
Papier aus verantwortungsvollen Quellen
Paper from responsible sources
FSC® C105338

If you have any concerns about our products,
you can contact us on
ProductSafety@springernature.com

In case Publisher is established outside the EU,
the EU authorized representative is:
Springer Nature Customer Service Center GmbH
Europaplatz 3, 69115 Heidelberg, Germany

Printed by Libri Plureos GmbH
in Hamburg, Germany